IMAGES
of America

AROUND DAMASCUS TOWNSHIP

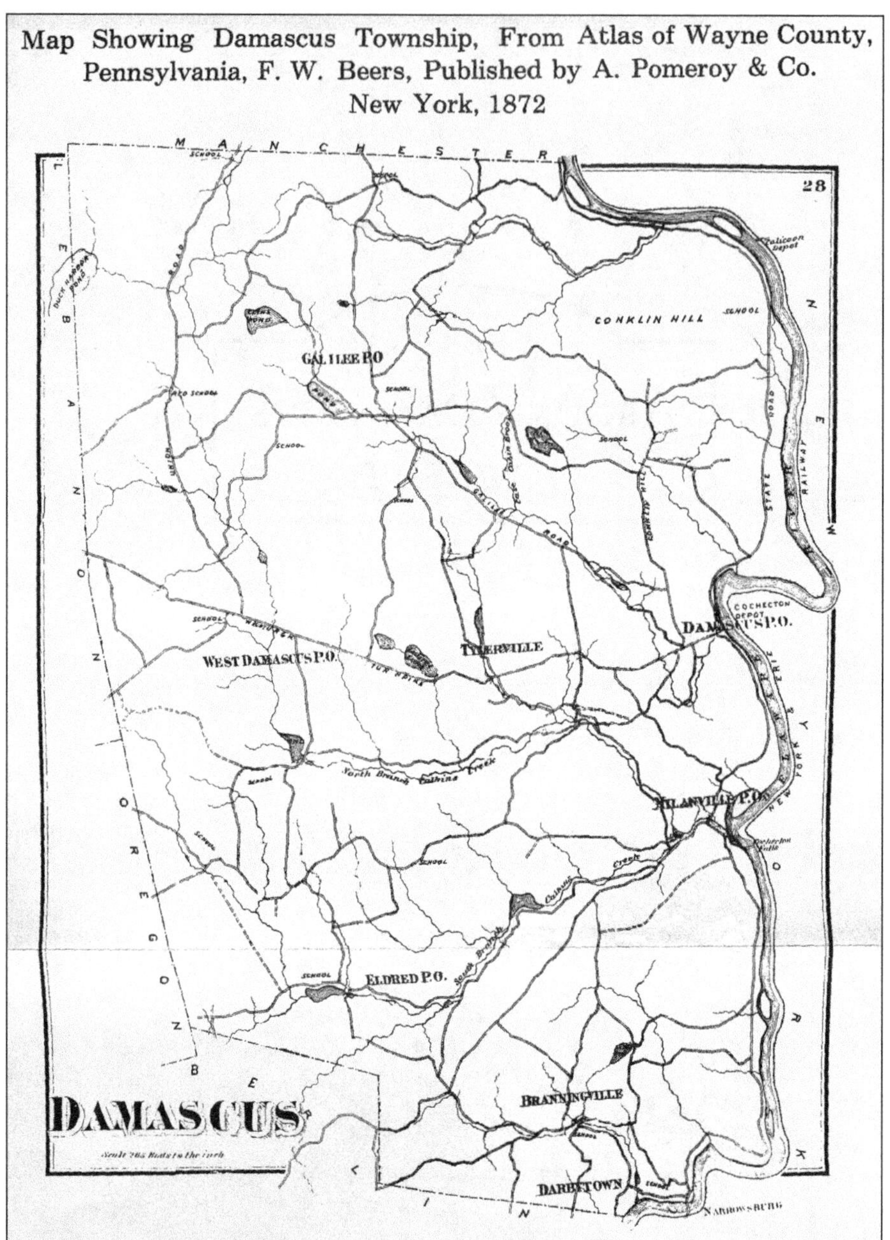

Map of Damascus Township. This map by F.W. Beers shows villages in the township. It was published in 1872 by A. Pomeroy and Company in New York. (Courtesy of the Damascus Township Historical Society.)

On the Cover: In the 1700s, the Bush and Tyler families were melded together by marriage and established the Bush's Glen sawmill north of Damascus near the Delaware River. Taking in the solitude, beauty, and remoteness of the glen, many would gather here for picnics and events. (Courtesy of the Damascus Township Historical Society.)

IMAGES
of America

AROUND DAMASCUS TOWNSHIP

Barbara Davis Dexter

ISBN 978-1-5316-6230-1

Published by Arcadia Publishing
Charleston, South Carolina

Library of Congress Control Number: 2011938700

For all general information, please contact Arcadia Publishing:
Telephone 843-853-2070
Fax 843-853-0044
E-mail sales@arcadiapublishing.com
For customer service and orders:
Toll-Free 1-888-313-2665

Visit us on the Internet at www.arcadiapublishing.com

This is dedicated to all Damascus Township history makers of long ago, and to those who cherish it today.

Contents

ACKNOWLEDGMENTS

Any major undertaking surely involves more than one individual. Whether they lend an opinion, quiet support, or an active role, each is important to the success of the project. I would like to thank the following people for all they have contributed.

To my father, Frank Davis, and grandfather John Davis for their work ethic and inspiring stories of life here long ago, I am forever grateful.

Many thanks to my husband, Schuyler, who picked up the slack and supported me with his enthusiasm and reassurance.

Valued are the personnel of Arcadia Publishing for their expertise, and in particular Darcy Mahan, who was always available with answers and to give guidance.

Special thanks to everyone who opened their precious family albums and shared their time: Helen Adams, George Banta, Doris Bryant, Karl Canfield, Kathy Card, Ivan Davis, Joe Davis, Leona Rolston-Dean, Rocco DeGori, Helen Dexter, Marie Diehl, Virginia Dutton, Eloise Fasshauer, Roger Flederbach, Tessie Gillow, Fred Haase, Harold Hawley, Opal Hocker, Evelyn Jones, Lynda Marks, Evelyn McGowan, Lorraine McGrath, Tammy Miller, Earl Mohn, Albert Noble, Hazel Olver, Sue Riefler, Brian Rutledge, Jacque Rutledge, Myra Schnakenberg, Grant Sheard, Josh Sheard, Ralph Smith, Donald Wood, and others who so eagerly took part in this venture.

Appreciated are the willingness and cooperation of many local historical societies and their staff. Especially noted are Roger Swendsen of the Damascus Township Historical Society, Carol MacMaster of the Equinunk Historical Society, Grace Johansen of the Tusten Historical Society, and Gloria McCullough of the Wayne County Historical Society.

Around Damascus Township has truly been a community effort, documenting the history of our rural life as no other book has done before. Sincere thanks to all.

INTRODUCTION

Cushetunk, founded by Daniel Skinner and Moses Thomas in 1755, was the first settlement on the upper Delaware River. Other families soon followed, and they made their homes in the river valley. While the Lenni Lenape Indians were content to share the land, the pioneers heard of onslaughts in other areas, so they built a fort for protection in the event of a raid.

Although William Penn came to the New World in 1682 to take possession of land granted by a charter of King Charles II, he believed the true owners to be the Indians who already lived there. Penn made a treaty with them, which he kept, and there was peace. However, local tribes and the Six Nations had disputes over this land ownership, which in turn caused turmoil between the white man and Indians.

John Penn, William Penn's grandson, arrived in 1763 as lieutenant governor in an effort to secure ownership of the land. His methods were not diplomatic but barbaric, having placed a bounty on the Indians. Within a year, the Indians were ready for peace. By 1768, a treaty resulted in the Six Nations selling the disputed land. This did not, however, give peaceful results. Encouraged by Great Britain and under their influence, the Indians continued fierce battles with the white man, both on a large and small scale.

After learning of the Indian massacre in 1778 at the Wyoming Valley area of Pennsylvania, some in the Delaware River valley feared for their safety and fled. Among those at Cushetunk and nearby who remained were John Land, Nathan Mitchell, Bezaleel Tyler, and Moses Thomas. The next year, some of them would meet their death at the Battle of Minisink, downriver from the fort. Iroquois raiders led by Joseph Brant, a Mohawk chief and British captain, mightily defeated 120 armed colonists.

With the end of the Revolutionary War in 1783, peace gradually returned and other villages were established in the Delaware Valley area. From the sacrifices of the first families and their fervent desire to settle in this untamed territory, Damascus Township was formed in 1798.

Even after other townships were augmented with land from Damascus Township, it is yet the largest in Wayne County. The 1800 population of 145 was comprised of 63 children younger than 16 years of age, 12 adults over the age of 45, and 70 of an age in between. Over 200 years have passed since then, and the township remains quite rural with the census in 2000 totaling just over 3,600.

Most early villages in the area were located next to creeks, lakes, or the river. With the vastness of virgin forests, lumbering was the primary industry, which required an ample water supply for manufacturing and transporting its goods. Dams were built, providing energy to run sawmills and gristmills. Raftsmen lashed their logs together and waited for the springtime snow and rain runoff, which raised the river level so they could steer their rafts to market at Bordentown and Trenton, New Jersey. Shipbuilders from Philadelphia would scout the land for the best timbers to be sent down the Delaware River for ship masts. In winter, ice would be harvested in blocks upwards of 300 pounds to be used for cooling perishables throughout the year.

In the Southern Tier chapter of *Around Damascus Township*, the locales of Darbytown, Atco, Ashland, Boyds Mills, Calkins, and Fallsdale are represented. Darbytown is situated across from Narrowsburg, New York. With Samuel Darby's prosperous tannery in operation by the mid-1800s, the area grew to be quite an active hub. Atco was originally known as Branningville, having been founded by the Branning family. At one time, there were blacksmiths, a gristmill, a store, and school established in the hamlet. Ashland was more farmland than industry, and had its own school. Boyds Mills, originally known as Eldred, is located on Calkins Creek and named after Thomas Boyd. Mills there were successful, and the village flourished. Calkins is the only one in this grouping that had a church, a grange, and a two-room schoolhouse. Fallsdale, with its dam and pond, touted a creamery, blacksmith shop, and sawmills.

Milanville, site of the original fort at Cushetunk, is the subject of the second chapter. Its history of perseverance and survival helped clear the way for settlement of the surrounding areas. Rafting, mills, a tannery, and acid factories were its earliest industries. The Erie Railroad, located on the New York side of the river, began operating in the mid-1800s and was a boon to local towns. In 1901, after the river bridge at Milanville was completed, a train depot, creamery, and hotel were built on the other side, and the locale became known as Skinner's Falls.

Damascus was part of a land purchase in 1796 by Thomas Shields, a wealthy jeweler from Philadelphia. He envisioned a town with streets, laid out lots for a church and school, and built gristmills and sawmills. In 1819, the first covered bridge was constructed across the river to Cochecton, New York, with the tollhouse located on the Pennsylvania side. The bridge, connecting the Newburgh–Cochecton Turnpike with the Cochecton–Great Bend Turnpike, made risky ferry and scow crossings a rarity.

The next chapter concerns the village of Tyler Hill, known as Tylerville prior to 1878, when the post office changed the name. In 1828, Israel Tyler had purchased his first land parcel of nine acres containing a sawmill and planing mill. Around 1850, he bought a large tract that included the reservoir lot. From this location, logs, lumber, and products, such as banister material and saddletrees, were sent downriver on his rafts.

The Northern Tier chapter highlights Conklin Hill, Galilee, Abrahamsville, Rutledgedale, and Lookout. Conklin Hill was set back from the river valley, and had a church and school. Early Galilee was a bustling village with stores, a barrel factory, cheese factory, creamery, church, school, grange, chicken coops, farms, and cattle drives down the main street. A church and school were located in Abrahamsville, primarily a dairy and chicken farming community. First known as Rutledgetown or Rutledgeville, the area of Rutledgedale is a sprawling region that had a school, mills, post office, and store at one time. Named for Alexander Rutledge, the hamlet was home to his 11 children and their offspring. Lookout had also been a village with a flurry of activity. With Joel Hill's business enterprises, the burg at one time took on the name of Hilltown. The existence of lumber mills, an acid factory, church, store, and school were all due to his endeavors.

The last chapter, Hobbies and Recreation, is about the lighter side of life. Much of history deals with facts and figures, and so it is easy to forget that celebrations and fun were a very important part of life, then and now. With the variety of recreation represented, readers will surely relate to some photographs with a smile.

The era when the original settlers and immediate generations thereafter lived was the most difficult time. They had to cope with the elements of nature, strangeness of the land and natives, uncertainty of health and longevity, and unending toil. From that first settlement through the thriving industrial age to the sleepy little villages today, *Around Damascus Township* is intended to enhance one's understanding of life in this area.

One

The Southern Tier

Darbytown on the River. The silence of leisure time on the Delaware River in this c. 1930 photograph is in stark contrast to the activity of long ago. Shipbuilders from Philadelphia came to this location in 1786 looking for timber to be used as tall, sturdy masts. Lore has it that with the help of local lumbermen, a sturdy pine was felled on the mountainside near Peggy Runway and then floated downriver to Philadelphia for use as one of the masts of the USS *Constitution*, or "Old Ironsides." An article dated April 21, 1939, in the *Delaware Valley News* states, "as early as 1801 there was a small settlement opposite Narrowsburg, New York. The tannery was built in 1851 by Horton and Company. Samuel Nelson Darby acquired the property and operated it until 1876 when the supply of hemlock bark ran out." (Courtesy of William and Eileen Dexter Miller.)

DARBYTOWN-NARROWSBURG COVERED BRIDGE. In 1810, the New York Legislature approved a charter for the first bridge across the narrows at Big Eddy. The charter belonged to the Narrowsburg Bridge Company with Jeremiah Lillie, Jonathan Dexter II, Chauncy Belknap, Thomas Belknap, Samuel F. Jones, William A. Thompson, William W. Sackett, Samuel Preston, and Francis Crawford as incorporators. That bridge was destroyed by a flood and replaced with the covered bridge pictured above. On the river is a group of 11 men harvesting ice from the Delaware River.

Although a horse-drawn snow plane could be used to remove the snow from the ice, shovels are being utilized in this photograph. The two men on the left are using breaking spades, while the one next to them is handling a needle bar. The apparatus, seen hooked up to the single horse in the background, appears to be a cutter, which scored the ice about two-inches deep. After harvesting, the ice would then be kept covered in sawdust in ice houses for use throughout the year. (Courtesy of the Tusten Historical Society.)

NEW YORK–PENNSYLVANIA TOLL BRIDGE. In 1899, the Narrowsburg Bridge Company hired out the work to build an iron suspension bridge. Tolls were collected until 1927, when free passage went into effect. In 1953, that bridge was replaced with the one currently in use. The sign on the bridge warns of a $10 fine for driving "faster than a walk." (Courtesy of the Tusten Historical Society.)

WEST POINT HOUSE. Around the 1900s and for many years after, this boardinghouse was owned by Thomas and Alice Lounsbury Dexter. Accommodations were for 30 guests at $6 per week and featured a pavilion for dancing. A 1906 advertisement stated, "One-half mile from station, conveyance free; bowling." (Courtesy of the Tusten Historical Society.)

Yatho Soda Bottling. About 1890, Charles J. Yatho started a soda bottling company on the edge of Darbytown and delivered to local areas with a horse and wagon. He obtained the skills necessary for such a venture during his time spent in the city. Years later, around 1920, he would deliver with a motorized Dodge wagon to Barryville and Port Jervis. Yatho's ginger ale was a favorite of Highland Lake resorts in New York. Upon his death in 1932, his wife, Mary, his daughter Marian, and her husband, Floyd Gillette, took up the business. The mountain spring from which he bottled water and made carbonated soda was located across the road, just opposite his home-operated business. (Above, courtesy of George Banta; below, courtesy of the Tusten Historical Society.)

Peggy Runway Lodge. In July 1929, Frank and Ruth Behling opened their hotel for its first season. William Heinz, the architect, designed the lodge in an English Tudor style, with 13 guest rooms on the second floor. Fred Beck, a Lackawaxen mason, used stones from the old Darbytown tannery foundation for the double fireplace. In the early days of operation, most guests arrived by train at Narrowsburg, especially when gasoline was rationed during wartime. (Courtesy of Grace Johansen.)

Atco School. On December 17, 1881, schoolhouse number five in Branningville (Atco) was destroyed by a fire. The Darbytown school, which had been closed for a year, reopened temporarily to accommodate the displaced pupils. Pictured from left to right at the Atco one-room schoolhouse are (first row) Emma Buckingham, Toots Willand, Louisa Dexter, Walter Terry, Marie Steffens, Olga Lilholt, Evelyn Barkley, Marion Yatho, and Margaret Horst; (second row) Roy Horst, Hap Lilholt, Brink Man, Thelma Lilholt, Minnie Steffens, Kenneth Smith, Carl Buckingham, and in the back, teacher Florence Sheard. Meda Reynolds was the last teacher when the school closed in 1947. (Courtesy of Rodger Flederbach.)

The Mill House. This large building, constructed in 1852 by John Branning, was to be used as a gristmill. In 1915, the Delaware Valley Grange 1564 purchased it. The grange in Atco, which was comprised of 31 charter members, had been organized two years earlier. Serving as its first master was L.S. Barkley. (Courtesy of Roger Swendsen.)

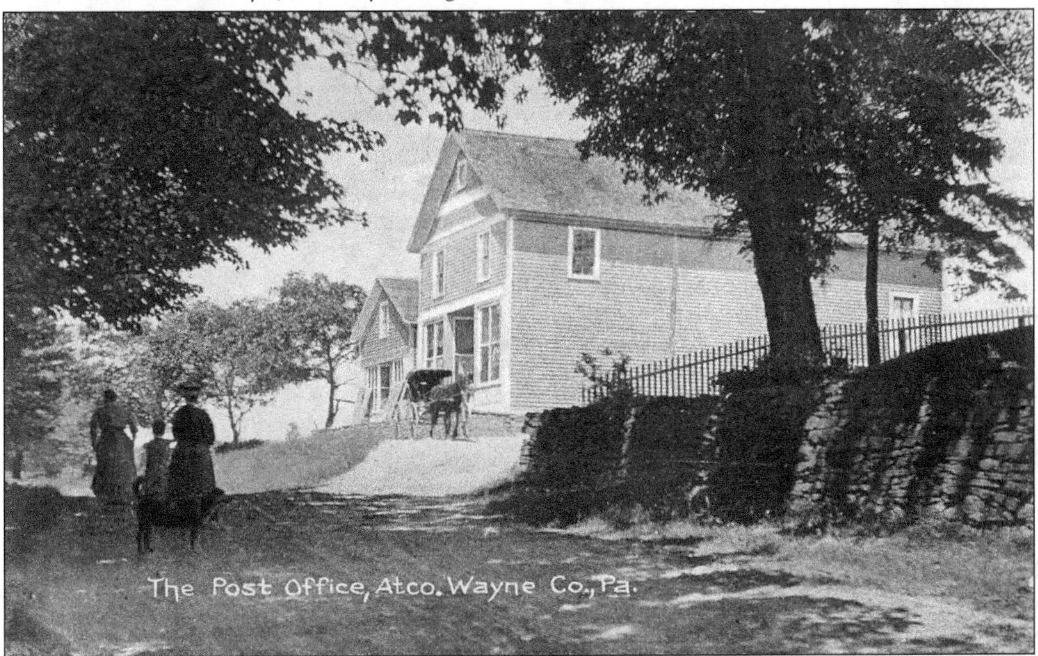

Atco Post Office. When William B. Guinnip was a member of the state assembly from 1884 to 1885, he secured the establishment of a post office for his location. Branningville, as the area was originally known, was too long a name, according to O.R. Packard. Packard purchased the Branning estate, renamed the town Atco, and was its first postmaster. The name Atco was taken from a town in New Jersey. (Courtesy of Lorraine McGrath.)

Threshing the Fields. Many farms did not have every piece of equipment necessary to complete the various agricultural chores. At this farm, Russell Dexter of Ashland is pictured overseeing workers using his horse-drawn thresher. He traveled throughout northeastern Pennsylvania and nearby New York hiring out his threshing services in the late 1890s and early 1900s. (Courtesy of the Dexter family.)

Young Family. In this c. 1918 portrait are descendants of local founding fathers Aaron Young and Moses Thomas, a Revolutionary War patriot. From left to right are (first row) Miles, Angenette (née Phillips), Wilmot, Elsie, Junius, and Horace; (second row) Elizabeth, Laura, Margaret, Hazel, and Warren. (Courtesy of the Young family.)

Flight School. Robert Young is shown suiting up for instruction at the Navy flight school in Michigan. After training with the "Yellow Peril" biplane, he served his country during World War II. (Courtesy of the Young family.)

Patriotic Youth Group. A local organization, similar to the Boy Scouts of America, would assemble youngsters for patriotic recreation and instruction. Among the group are three Ashland brothers, Herbert, Robert, and Richard Young, children of Horace and Anna Hocker Young. (Courtesy of the Young family.)

Ashland School
Damascus Twp., Wayne Co., Pa.
April 10, 1914.

Myrtle I. Reynolds, Teacher.

Superintendent of Schools, J. J. Koehler.

PUPILS

Horace Young
Warren Young
Floyd Reynolds
Grant Reynolds
Louis Reynolds
Herman Mohn
Raymond Mohn
Florence Reynolds
Luella Reynolds
Elizabeth Young
Laura Young
Ethel Anderson
Sophia Dexter
Charlie Dexter

Ashland School Memento. Photographs were not always taken every year at school, so a memento would be handed out instead. In April 1914, the Ashland School had a total of 14 pupils, with Myrtle I. Reynolds as teacher. Students listed are Horace Young, Warren Young, Floyd Reynolds, Grant Reynolds, Louis Reynolds, Herman Mohn, Raymond Mohn, Florence Reynolds, Luella Reynolds, Elizabeth Young, Laura Young, Ethel Anderson, Sophia Dexter, and Charlie Dexter. The school closed in 1930 with Harriet Allen as the last teacher. (Courtesy of Earl Mohn.)

Mitchell Family. On July 22, 1779, Capt. Bezaleel Tyler was killed at the Battle of Minisink. Ten weeks later, his wife, Abigail (née Calkins), gave birth to a girl, also named Abigail. She married Joseph Mitchell at age 19, and they had a family of 10 children. Nine are pictured here, from left to right, as follows: (first row) David, Elizabeth Jane (wife of Thomas Y. Boyd), Mary Ann (wife of John Y. Tyler), Elmira (wife of Joseph Wood), Violetta (wife of William H. Gavitt), and Nathan; (second row) Albert, Oliver, and Bateman. (Courtesy of Jonathan Sheard.)

Wood Family at Home. The Wood family was established in the Boyds Mills area in 1837 with the arrival of Stephen Wood. His son Joseph Wood was in partnership with Thomas Y. Boyd, purchasing the Willis and Tymerson sawmill in 1867. Descendants are pictured here from left to right (first row) Raymond, Flora, Augusta (Noble), Clark Sr., Lucille (baby), Kenneth, and Eunice; (second row) Amy and Clark Jr. (Courtesy of Donald Wood.)

Boyd Family. Descendants of Thomas Y. and Elizabeth Jane Mitchell Boyd are pictured in this late 1890s or early 1900s photograph. From left to right are (first row) Caroline Clark Myers, Elizabeth Jane Mitchell Boyd, Myron Boyd, Edna Boyd, John Ellis Noble, and Boyd Clark; (second row) Robert Boyd, Mary Noble Branning, Russell Clark, Anita Clark Cotner, Clarence Noble, and David Boyd. (Courtesy of Jonathan Sheard.)

Boyds Mills in 1885. At the time this photograph was taken, the village was known as Boyds Mills. It would undergo another name change in two years, before the permanent name of Boyds Mills would be established in 1889. The beginning of the town took place many years prior. In 1843, Hiram Willis and Truman Tymerson built a sawmill there, which was purchased by Joseph

Wood and Thomas Y. Boyd in 1867. Upon the passing of Wood in 1877, Boyd became the sole owner of the mill. In March 1889, at the age of 66, Boyd passed on, and two months later, his widow became postmistress of the village. (Courtesy of Eloise Noble Fasshauer.)

The Stone Arch Bridge of 1894. An agreement between Wayne County commissioners and Joseph Boyd stated, "For the sum of $480, all of said work to be completed by October 1894, and be done in a good and workman like manner." Commissioners listed on the document were J.B. Keen, Joel G. Hill, and W.E. Perkins. (Courtesy of Lorraine McGrath.)

August Smith Mill. Downstream from the village on Calkins Creek was a crudely built mill. Owned by August Smith, the mill employed many men and also used teams of horses and oxen. John Davis is pictured driving the team of horses, and at the wheel of the automobile is a member of the Boyd family. They had closed their mill business in 1913 but opened a car dealership in Honesdale. The license plate reads "Penna, 1913, dealer." (Courtesy of Eloise Noble Fasshauer.)

Boyds Mills School. One of the first schools in Eldred was an old block schoolhouse built around 1836 with Emily Stearns as teacher. In 1850, another schoolhouse was constructed. In this fall 1904 photograph, 22 pupils attended the Boyds Mills one-room schoolhouse. From left to right are (first row) Clark Wood, Merle Bogart, Arthur Decker, James Decker, Boyd Clark, Eugene Conklin, Coe Decker, Searle Wood, Russell Clark, Forrest Wood, Ralph Wall, Robert Boyd, Joseph Gavitt, and Millard Rice; (second row) Bertha Ostrander, Inez Decker, Anita Clark, and Flora Wood; (third row) Alfa Gavitt, Mary Rutledge, Amy Wood, and Dove Wilcox. In the back is teacher Ida Davey. (Courtesy of Donald Wood.)

Clarks at Home. Russell and Caroline Clark are seen in the parlor of their family home in Boyds Mills. After graduating from Bloomsburg State College in 1905, she instructed illiterate soldiers in the US Army throughout World War I. In 1912, she and Dr. Garry Cleveland Myers married. Years later, they founded the magazine *Highlights for Children*. (Courtesy of Eloise Noble Fasshauer.)

Expeditionary Force. Pvt. Clarence A. Noble was a member of the 25th Balloon Corps in France from 1917 to 1919. As part of the American Expeditionary Force, he did surveillance over enemy lines at night from an observation balloon. Noble was in Paris on November 11, 1918, for the signing of the armistice, and returned home in 1919. (Courtesy of Eloise Noble Fasshauer.)

Sheard's General Store. As was the custom of rural general stores, much more than just goods were supplied. It was a community gathering spot where one could meet the neighbors and spend some time conversing about the latest news. Pictured in 1941 on the porch from left to right are (first row) William Swendsen, Grant Sheard, and Russell Swendsen; (second row) Eva Taylor, Elizabeth Warren holding Garry Sheard, Edna Swendsen, Floyd Sheard (owner), and Dr. Garry Myers. (Courtesy of Grant Sheard.)

Transporting Milk. The winter of 1945 was a memorable one. According to local Floyd Sheard, "It snowed and blowed all winter so bad that there was no school for six weeks. We had to make paths anywhere we could just to get around. The farmers would bring milk cans on horse drawn sleighs to Boyds Mills, and then Al Henderson Sr. would transport it to the Narrowsburg Creamery and bring back feed and such." (Courtesy of Grant Sheard.)

Temporary Bridge. After the flood of 1952 took out the stone arched bridge, work began on a temporary one in Boyds Mills. Known for his abilities and work ethic, John H. Davis was hired, at the age of 72, as foreman of the project. Among the workers pictured are John H. Davis, Floyd Sheard, Peter Wood, and Grant Sheard driving the tractor. (Courtesy of Grant Sheard.)

Boyds Mills Post Office. Established as Eldred in 1837, the village's first postmaster was James Smith. The name was changed to Boyds Mills in 1880, with John Orr as postmaster. The village had yet another name change to Astoria in 1887, with Christopher T. Tegeler serving. In 1889, the name was changed permanently back to Boyds Mills upon Elizabeth J. Boyd serving as postmistress. Floyd Sheard served as acting postmaster from September 1, 1933, until the office's closing. Pictured with Floyd Sheard (left) is Caroline Myers on its closing day, August 17, 1973. (Courtesy of Grant Sheard.)

Boyds Mills General Store. Originally built by Tymerson and Willis in the early 1800s, the company store kept mill workers supplied with goods. Thomas Boyd obtained ownership of the store when he and Joseph Wood purchased the lumber business from Tymerson and Willis. Since then, the building was moved to its present location and saw a variety of owners, including the Boyd and Clark families, Jacob Lounsbury, William Tegeler, C.T. Tegeler, Claire Tegeler, Willard Dillmuth, and Floyd Sheard. It had also been used as a polling place for elections for a number of years. When the post office closed in 1973, so did the store. (Courtesy of Grant Sheard.)

Dairymen's League, Calkins Plant. Founded in 1907 in Orange County, New York, the Dairymen's League was one of America's first cooperatives. With the goals of fair pricing and a guaranteed market, the league began operating its own processing plants, such as this one in Calkins. (Courtesy of Hazel and Tom Olver.)

Maple Grove Farm. In the early 1900s, Calkins was the setting for this boardinghouse, owned by Silas and Mary Saunders Noble. The working farm took in boarders and featured gaslights, a telephone, and transportation from the rail station. In 1940, at the age of 75, and five years after becoming a widow, her son's family came to live with Mary Noble. When Samuel and Mildred Olver Noble moved in, they had eight children, with another being born later. (Courtesy of Virginia Noble Dutton.)

Allen Grove School. This one-room schoolhouse held sessions from 1902 until closing in 1920. The first teacher was Judson Noble, and the last was Alpha Crocker. In the early years of operation, the usual number of students attending was 12 or fewer. (Courtesy of Eloise Noble Fasshauer.)

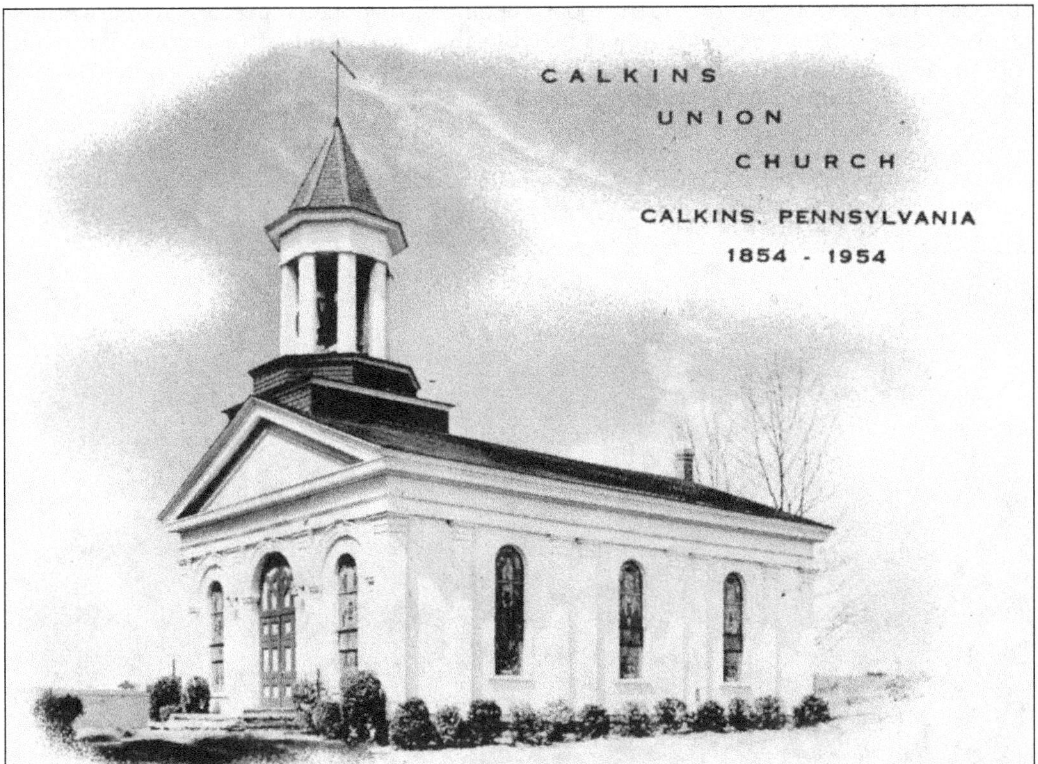

CALKINS
UNION
CHURCH
CALKINS, PENNSYLVANIA
1854 - 1954

The Union Church. Originally known as the Beech Union Church, it was erected in 1854 at a cost of $2,500 and funded by three denominations: Methodist, Baptist, and Presbyterian. Remodeling of the edifice was done in 1899 with additional help from the Disciple Society. Repair work to the steeple, which was damaged by lightning, resulted in a different shape. Instead of the previous round steeple and pillars, the new shape was a hexagon with square pillars. (Courtesy of Helen Dexter.)

Burcher Hill School. While most one-room schoolhouses had their water carried from a nearby spring, this one's water supply was from a barrel in the beaver pond at the bottom of the hill. Prior to the decline in number of pupils and its closure in 1916, the school had an attendance of 28 at its peak. (Courtesy of the Wayne County Historical Society.)

Calkins Consolidated. In 1922, a two-room schoolhouse was erected at Calkins to replace the one-room schoolhouses of Allen Grove, Burcher Hill, Fallsdale, and Boyds Mills. It housed an average of 80 pupils in the 1930s, and in 1946, when attendance dropped to 25, it became a one-room schoolhouse, with the other portion of the building used for recreation. The school closed in the early 1960s, with Rosamond Sheard serving as teacher. (Courtesy of Doris Bryant.)

Village of Fallsdale. The Fallsdale Creamery Company was situated at the base of the hill near the pond. Pictured to the right of the company is Joseph Sheard's blacksmith shop. On December 23, 1913, J.T. Bradley wrote, "Joseph Sheard of Fallsdale was recently granted a patent on a plumber's torch that is said to eclipse any other on the market." Located just a short way from the Lovelass Mill was another sawyer, Jabez Stearns, who turned his mill over to his son. The creamery closed in June 1914, and the post office, which was established in 1897, was discontinued in June 1921. (Postcard courtesy of Roger Swendsen.)

Rolston Homestead. Pictured by Laila and Charles Rolston's home in late winter 1912 are, from left to right, Louise Gaston Rolston holding daughter Mabel, Walter Swendsen, Laila Lewis Rolston, Wilhelmina Rolston, Charles Rolston, Ethel Rolston Stephens holding daughter Lucille, and Gertrude Rolston Swendsen holding son Ivan. This house later became the John Stephens farmhouse, with Merlin Stephens occupying it until the late 1980s. (Courtesy of Roger Swendsen.)

Upper and Lower Sawmills at Fallsdale. In the early part of 1800s, John Leonard purchased a large tract of land where Fallsdale is now. He built a sawmill below the falls on the north branch of Calkins Creek and operated it for 10 years. Eli Beach bought out Leonard, and in 1850, he passed it into the ownership of Joseph Wood and Thomas Boyd. In 1859, Isaac Lovelass bought a half interest in the business, and in 1880, he became sole owner. A dam was built during this time above the falls, forming a pond that supplied water for yet another lumber mill. (Courtesy of the Damascus Township Historical Society.)

Two

MILANVILLE

THE VILLAGE OF MILANVILLE. In this c. 1900 photograph, the chemical company is seen on the right. It was one of the largest in the country, at 80 feet wide and 260 feet long. The company store is just in front of it, a little to the right. Eli Beach, once owner of the factory, had the home with porches at the base of the hill, left of the road. Up the hill on the left above the barn, Volney Skinner's house can be seen, which had been used during the Civil War to house southern prisoners being transported to Elmira, New York. Below the convergence of the mountains in the background, the one-room schoolhouse can be seen. In the foreground is the mouth of Calkins Creek flowing into the Delaware River. (Courtesy of Lorraine McGrath.)

The Red House. John Land, a colonist, was imprisoned during the Revolutionary War, but returned to Cushetunk after his release. He then purchased back the 438-acre Land family farm from Joseph Thomas and Jonathan Dexter. He married Lillie Skinner, and in 1796 built the "red house," as it is known today. (Courtesy of Helen Dexter.)

The Old Bridge. In the late 1800s, this bridge was located at the junction of the north and south branches of Calkins Creek at Milanville. Note the differences in construction of the spans. Either one needed replacing, or both were not built under the direction of the same person. (Courtesy of the Card family.)

Skinner's Sawmill. Brothers Milton and Volney Skinner began building their sawmill on January 1882 on the banks of the Delaware River, just above Cochecton Falls. Boilers, fired with sawdust, ran the 30-horsepower steam engine, giving the mill a capacity of 10,000 board feet per day. Many locals were employed there, as the business had a planer, shingle, and lath mill. Logs from Pennsylvania were brought to the mill, and a scow that the Skinners operated transported timber from New York. Often, finished lumber would be rafted down the river to Bordentown, Trenton, and Philadelphia. In 1905, an 80-acre parcel along with the mill was sold to C.H. Rexford of Galeton, Pennsylvania. Rexford's mill, seen below, was torn down in 1918. (Above, courtesy of Lorraine McGrath; below, courtesy of Helen Dexter.)

Store at the Unnamed Settlement. J.T. Bradley is quoted, "When work was temporarily stopped on the Erie Railroad, and Chauncy Thomas, a storekeeper at Deposit, NY, had no customers, his stock of goods were loaded on a raft and moved to a nameless settlement (Milanville) in 1835 down river. One day when Col. Calvin Skinner was in the settlement's store and Thomas was signing shinplasters, he remarked to Skinner that to write the name of Damascus upon them seemed remote. Col. Skinner suggested that he write 'Milanville.' This was done and the name yet sticks. The suggestion was made because the colonel was a great reader, and admirer of Napoleon's Decree issued at Milan, so he added the suffix, -ville." Pictured c. 1900 is the original store. (Photograph courtesy of Helen Dexter.)

Bridge across Calkins Creek. This c. 1900 Milanville bridge with metal arches had planking as the base in order to drive across. Sturdiness was a main feature of the bridge, since heavily laden wagons of charcoal, or sleighs loaded with ice blocks in winter, would often travel that route. In the scene below, the general store is on the right, the front steps of the post office reach the road halfway up the hill, and at the top of the hill is the schoolhouse. (Above, courtesy of Lorraine McGrath; below, courtesy of the Wayne County Historical Society.)

THE MILANVILLE CHEMICAL COMPANY. In the early 1900s, acid factories were in demand and were located where forests were rich in hardwoods. Products from acid wood included acetate of lime, wood alcohol (methanol), and charcoal. These chemicals were important during World War I when they were used in explosives. This factory employed 50 or more men. The labor of wood cutters, teamsters, and factory workers kept the business going 24 hours a day. The factory itself needed 10 to 15 men, in twice daily shifts. Employees pictured here are, from left to right, John Scherer, Al Jay, Marvin Conklin, and two unidentified. (Courtesy of Schuyler Dexter.)

RIVER BRIDGE AT MILANVILLE. The Milanville Bridge Company initiated the construction of this bridge. After legal wrangling caused by owners of the nearby Damascus-Cochecton and Darbytown-Narrowsburg bridges, a charter was issued in 1901. The tollhouse was located on the Pennsylvania side of the bridge with the price of 25¢ for a horse and wagon to cross. The last toll tender was Albro Dexter in 1928, at which time the bridge commission purchased the bridge and crossings were made toll-free. (Courtesy of Helen Dexter.)

THE VOLNEY SKINNER HOUSE. Built in the early 1800s, the house had first been used as a hotel for raftsmen and lumberjacks, and then in the early 1900s accommodated tourists traveling by rail. Bessie Skinner was the last of the family to occupy the home in 1972. She was the daughter of Volney Skinner and great-great-granddaughter of Daniel Skinner, one of the original settlers of the area. (Courtesy of the Wayne County Historical Society.)

VILLAGE HISTORIAN. Bessie Skinner was a direct descendant of Daniel Skinner, Lord High Admiral of the Delaware River. Born in 1877, she was the daughter of Volney and Sarah Lillie Skinner and lived to be the age of 95. With her rich heritage and attention to recording historical details, she was honored when her notes and published articles were accepted into the Smithsonian Institute. (Courtesy of the Smithsonian Institute and the Wayne County Historical Society.)

Skinner's Falls Hotel. In 1908, John E. Skinner set out to build a hotel across from Milanville and above the falls in New York state. In the early photograph above, the hotel has bare bones beginnings, and appears to be a raftsmen stopover. In 1914, he sold the property, and over the years the Skinner's Falls Hotel had grown to a prosperous establishment, as seen in the photograph below. In the mid-1900s, Frank Plessel owned the business, and square and round dances were held in the ballroom in the back. Local families looked forward to meeting their neighbors at the Saturday-night dances. Buddy Howe's Orchestra usually provided music. (Above, courtesy of the Tusten Historical Society; below, courtesy of Lorraine McGrath.)

SKINNER'S FALLS DEPOT AND MILK STATION. In February 1917, The Public Service Commission ordered the Erie Railroad to "Erect a suitable depot at Skinner's Falls by June 1, and maintain it properly thereafter. The business done at that place warrants such an advancement." In July 1956, a round trip fare cost $6.75 from Skinner's Falls to New York City. The Fulboam Dairy operated the milk receiving station at Skinner's Falls in the early 1900s. (Courtesy of Helen Dexter.)

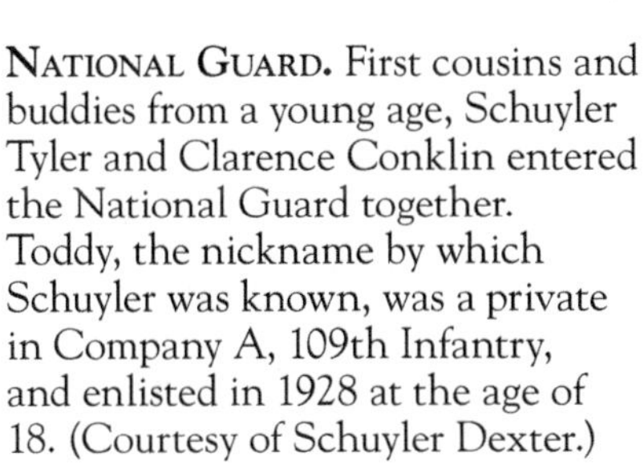

NATIONAL GUARD. First cousins and buddies from a young age, Schuyler Tyler and Clarence Conklin entered the National Guard together. Toddy, the nickname by which Schuyler was known, was a private in Company A, 109th Infantry, and enlisted in 1928 at the age of 18. (Courtesy of Schuyler Dexter.)

Cochecton Falls Farm, c. 1900. Located at the foot of present-day Skinner's Falls, this boardinghouse was owned by Louis J. Hocker. The three-story building featured "gas lighting, a bath on the second floor, telegraph office convenient, telephone in house, conveyance and baggage free one way from Skinner's Station, with rates eight to ten dollars per week." The brochure stated that meals were "bountifully supplied with the best that the country and season afford." A complete farm experience was offered to vacationers, including hayrides. (Courtesy of Ivan Davis)

COCHECTON FALLS BUNGALOWS. Renting one of the cabins below what are now known as Skinner's Falls was an attractive alternative to boardinghouses. Being situated right on the banks of the Delaware River, one could swim, boat, or just enjoy the view from the porch. In Erie Railroad's 1929 *Vacation Suggestions*, there were 12 bungalows available, all owned by A.J. Thomas. In the 1950s and 1960s, many of the bungalows were rented by the same families year after year for the entire summer season. Looking north, up the river, the early-1900s Fulboam Creamery and Skinner's Switch train depot can be seen. (Courtesy of Lorraine McGrath.)

Working at the Milanville Acid Factory. The wood distillery was in operation from 1898 to 1920, producing charcoal, wood alcohol, and chemicals used in explosives during World War I. There were 14 pairs of retorts, where the wood was loaded to be burned (charred). One of the factory's four teams of horses is seen here with a wagon of bagged charcoal ready to be shipped by rail. At the side is the store, with weigh scales under the canopy. (Courtesy of Schuyler Dexter.)

Gristmill Dam. Many stories about the beginnings of the gristmill exist. One possible builder was John Land around 1800, while it may have also been built by William Bonesteel of Damascus, or Jonathan Yerkes of Philadelphia. Undisputable is the fact that it had been an operating mill for over 150 years, the last miller being Orville Kays, who lived just up the hill. When high water took out the gristmill's dam in 1933, milling ceased. (Courtesy of Ivan Davis.)

Early School Days. Pictured attending the Milanville one-room schoolhouse around 1937 are, from left to right, as follows: (first row) Norman Vannatta, Marvin Spiess, Maynard Dexter, Irwin Cramer, David Illman, Lois Wood, Raymond Cramer, Irene Tyler, Betty Vannatta, Mildred Wells, and Richard Spry; (second row) Leonard Rolston, Theodore Wells, Robert Dexter, Earl Mohn, Robert Spry, Lucille Scherer, Lillian Dexter, Marian Davis, Juanita Hocker, Joan Cupido, and Shirley Foster; (back row) Vernon Davis, Donald Wells, Floyd Wells, Albert Jay, Orson Davis, William Whent, William Rooney, and Percy Tyler. (Courtesy of Earl Mohn.)

Haase Fruit Growers. Herman Haase became an accomplished orchardist, having earned the award of "Master Farmer" in 1938 by the *Pennsylvania Farmer*. Pictured in front of the packinghouse in this 1939 photograph from left to right are Emma Haase, Sophie Heins, Herman Haase, Florence Nichols, Dorothy Kleffman, two unidentified, two unidentified gentlemen from Penn State University, Alice Erhardt, Edward Steffens, Alfred Haase, Vincent Miller, unidentified, and Malvin Turner. (Courtesy of the Haase family.)

Spraying the Orchards. Since the individual orchards of Herman Haase and Henry Heins were near each other, it was mutually beneficial to share equipment and manpower. Using this horse-drawn sprayer in 1921 are, from left to right, Herman Huber, Haase, and Heins. Their working relationship continued until a feud later on made them part ways. In 1956, Haase's son Alfred purchased Heins's orchards. (Courtesy of the Haase family.)

OPEN HOUSE, FALL 1940. Anywhere from 800 to 1,000 visitors would come to the Haase packinghouse on the weekend of open house. Between 1911, with four acres and a few trees, to 1940, with many acres and over 38 varieties of apples, Herman Haase had become a well-known and respected orchardist. In addition to apples, he also grew sweet cherries, persimmons, and a variety of nuts. (Courtesy of the Haase family.)

THE PACKINGHOUSE. With the Haase orchards producing an abundant crop, it became necessary in 1923 to convert the property's barn into a packinghouse. In 1930, Herman Haase shipped 3,500 bushels of apples to Germany, where they were scarce that year. Pictured here in 1939 in the newly constructed cold storage area are, from left to right, Herman Huber Sr., Herman Huber Jr., Alfred Haase, Vincent Miller (on lift), and Herman Haase. (Courtesy of the Haase family.)

TELEGRAPH CREW. There was no need for hotel accommodations while these workers were constructing telegraph lines in the early 1930s, since the Pennsylvania railcar behind them was their sleeping quarters. It was equipped with berths and necessities for life on the job. When lines in an area were completed, the car and crew were taken to the next location and parked. The Davis brothers from Milanville are pictured here near Philadelphia. Frank is on the left, and Charles is on the far right. (Courtesy of Joseph Davis.)

SERVING DURING WORLD WAR II. James Card, the son of Norman and Eliza Orr Card, was a Merchant Marine during World War II, from 1944 to 1946. He returned home in 1947 to work the family farm with his wife, Frances (née Olver), after his father had taken ill. (Courtesy of the Card family.)

Milanville One-Room School. In 1957, this school had a total of 33 students in grades one through six. At the front was a long bench where children received class instruction from the teacher, while older children would be either helping younger ones or doing their own lessons. Seen at the back of the room (left to right) is the girls' cloak room, the teacher's room, an alcove with library and bell rope, a washroom with lunch boxes and water crock, and the boys' cloak room. Seated from front to back are (far left row) Barbara Davis, Joseph Davis, and Schuyler Dexter; (second row from left) Nancy Tyler, Terrence Branning, Dale Hocker, Harold Lassley, and Douglas Dillmuth; (center row) Marilyn Friermuth, Philip Masker, Rodney Mohn, Anthony Smith, Gregory McClure, and Duane Gillette; (right row, seated in pairs) Duane Rolston and John Moser, Carol Tyler and Patricia Mohn, Nancy Heins and Lois Mott, Kathy Gavitt and Lynne Ropke, Lawrence Mohn, Patricia Moser and Maureen Jay, Susan Jay and Leona Rolston, and Edwin Tyler; (standing) Evelyn Tyler, Kathy Griemsman, Lillian Mott, Grace Lassley, Sally Branning, and teacher Bertha Finch. (Courtesy of Barbara Davis Dexter.)

'Til We Meet Again. Romaine Tyler is seen here wishing her brother-in-law John Tyler a safe tour of duty. He served with Patton's Third Army in France, Belgium, and Luxembourg. In 1945, he was wounded in action while fighting in Germany. (Courtesy of Schuyler Dexter.)

Flood by Gristmill. Milanville was hit hard by flooding in July 1955. The road had been severely washed out with no possibility of travel, as the raging waters of Calkins Creek swept away most of the bridge, leaving part of it hanging precariously. The old gristmill on the left, built around 1800, weathered the storm quite well, given its age. It had not been in operation as a mill since 1933, when high waters took the dam out. (Courtesy of the Tyler family.)

Calkins Creek Flood. Many people in Milanville were stranded until the flood waters receded to a somewhat safe level. Families that lived above the general store, as well as those on the road behind it, had to be rescued by boat. Bridges were washed away around the gristmill and store areas, and low-lying roads were in shambles. Unless there had been a dire emergency, travel was not necessary, for the flood occurred during a time when neighbors helped neighbors, people grew their own food, and put up enough in the larder for a good long while, just in case. This was just one of those cases, as many opened their doors to friends until they could return home. (Courtesy of the Tyler family.)

At the Church Picnic. Although Milanville had but one church—Methodist in faith—all of the community was invited to their annual summer picnic. After a potluck dinner, games of all sorts were played. Enjoying fellowship in this early 1940s photograph are, in no particular order, Gertrude Meisy, Anna Holl, Juanita Hocker, Mildred Lassley, Mary Brucher, Beulah Davis, Betty Davis, Julia Davis, Minnie Lassley, Myrtle Hocker, Mr. Meisy, Lizzie Heib, Edna Rolston, Mrs. Patsky, Blanche Gabel, and Anna Young. (Courtesy of Joseph Davis.)

Returning from World War II. A joyful reunion between first cousins Lucille Scherer and Arthur Kniesel took place at the homestead of his parents, Frederick and Sarah Scherer Kniesel, in Milanville. (Courtesy of Schuyler Dexter.)

Big Eddy Telephone Company. This c. 1890 building in Milanville was put to good use in more modern times as a warehouse. In the Big Eddy fleet is a new 1958 Chevrolet service truck, second from left, just purchased from the Jack Denny Chevrolet Company. Personnel in the photograph are, from left to right, Delmore Olver (wire chief), Ivan Davis (lineman), Elwin Davis (trouble man and installer),Walter Kellam (lineman), Charles Davis (construction foreman), Arthur N. Meyers (president), Alma Tyler Philo (plant accountant), and Willard Dillmuth (general manager). (Courtesy of Ivan Davis.)

Dams of Milanville. This dam, which has a walkway on top, was located below the confluence of the branches of Calkins Creek and was instrumental in providing water for the raceway. Walls of the raceway were hand-laid stone, with some walls still visible today. Logs would be transported through the water-filled raceway to the site of the original Skinner mill, located near the junction of the creek and the Delaware River. In the photograph below is a dam which supplied water to the acid factory for use in its manufacturing. (Left, courtesy of Schuyler and Barbara Dexter; below, courtesy of Helen Dexter.)

Milanville Dairy. Herman Esselman, pictured with his hobby of equines, started managing the creamery owned by Franklin Lakes Dairy of New Jersey in 1944. The granary seen in the background became a safe haven for his family during a nighttime flood in July 1955. Their home was located just a little east of the granary, and the Calkins Creek ran behind both. Esselman's family recalls, "The water was around my dad's waist when he took us from the house to the feed storage building. He then went to the creamery to check on things. At that point the water was neck high. He disappeared into the dark and we didn't know if we would ever see him again. In the morning he showed up, much to our relief." Damage seen here to the creamery and surroundings was extensive. (Courtesy of the Esselman family.)

Filling the Hay Mow. With wooden hay forks in hand, this crew is ready to start stocking the mow for use during the winter. The huge load of hay had been loaded by hand and again would be transferred by hand into the mow of the Arlie Tyler farm. Seen here atop the mound, left to right, are Mickey Gerkins, unidentified, and Percy Tyler. (Courtesy of Richard and Nettie Tyler Swendsen.)

Break Time. Even with the help of mechanical equipment, farm chores required much of the workers. Seen here taking a break from cutting corn are Fred Pingle, who is pouring refreshment for Percy Tyler, and John Arlie Tyler looking on. (Courtesy of the Tyler family.)

On Cushetunk Flats. The Sunday school is pictured gathering at the big tree on the river flats to hold their community picnic in 1935. The boys made the ice cream, lemonade was served from a milk can, and potted meat sandwiches were the main course. Assembled here, in no particular order, are Mabel Dexter, David Illman, Kay Card, Alma Tyler, Isabel Conklin, John Dexter, Mary Console, Leonard Rolston, Eleanor Rolston, Maryann Sherwood, George Hocker, Lavina Jay, Evelyn Tegeler, Otto Tegeler, Muriel Sherwood, Willard Hocker, Clarence Jay, Mary Hocker, Norman Card, Mrs. Kays, Madge Orr, Annie Kays, Katie Console, Cora Gordon, Fannie Hocker, Blanche Gabel, Liza Card, Edna Tyler, Ethel Illman, Ruth Tegeler, Mrs. Brucher, Ruth Orr, Leila Orr, Esther Stephens, Christine Gerken, Margaret Orr, Zilla Orr, Libby Sherwood, Adon Sherwood, Janet Hocker, Margaret Spry, and Fred Tegeler. (Courtesy of Lavina Jay Powell.)

MILANVILLE SAWMILL. Before Milton and Volney Skinner's mill was built at the eddy of the river, they had built a mill near Calkins Creek just above the Delaware River. The raceway they constructed, however, did not operate as anticipated, and the mill closed. (Courtesy of the Card family.)

DELAWARE VALLEY FARM. This large boardinghouse in Milanville was owned and operated by Edward H. and Nora Lounsbury Dexter. Upon retirement, they relocated to the Atco area. (Courtesy of Ivan Davis.)

On Furlough. The Korean War was in its third year when Ivan Davis was drafted into the US Army. After basic training, in March 1953 he spent nine days at home before shipping out to Seattle, Washington, then to Tokyo, Japan, and on to Korea. Finishing his tour of duty, he returned home in August 1954 and served five years in the reserves. (Courtesy of Ivan Davis.)

Heins' Hill Snow. In the late 1920s and early 1930s, before the area had snowplows, crews of workers would shovel the roads by hand. Drifts especially needed to be cleared, as horse drawn sleighs loaded with milk cans had to get through to the creamery. Pictured shoveling on Heins' Hill road from left to right are John Davis, unidentified, Charles W. Davis, and unidentified. (Courtesy of Joseph Davis.)

A MATCH MADE IN HEAVEN. Arthur Rolston is seen driving his perfectly matched team of workhorses in this c. 1930 photograph. The team was not only used on land, but on ice as well. Farriers would fit shoes with cleats, pins, or nails for traction on ice. While pulling a load of harvested ice off the Delaware River, Rolston's team broke through the surface. In an effort to swim free, the horses flailed, and their shoes badly lacerated their legs. The team perished at the river as a result of their injuries. (Courtesy of the Brucher and Rolston family.)

Milanville Episcopal Methodist Church. In the photograph above, the community is assembled to celebrate the laying of the foundation for the Milanville Episcopal Methodist Church. Prior to the village having a formal church building, services were held at the schoolhouse, as Methodism has had roots in the local area since the 1700s. The church was built in 1910 on land donated by the Skinner family, and had eight baptisms that first year. The last church services were held about 2005, and it is now a privately owned chapel. (Above, courtesy of the Brucher and Rolston family; below, courtesy of Lorraine McGrath.)

George and Mary Hartz Brucher. Observing a quiet time together are George and Mary Hartz Brucher. Their family of three sons, August, Leonard, and Carl, and two daughters, Lizzie and Edna, had been raised on a farm in Milanville. (Courtesy of the Brucher and Rolston family.)

Officers of the Ladies Aid. Church-related expenses were often met with financial help from the Ladies Aid group. Chicken and biscuit suppers as well as sales of baked goods and handiwork raised funds to augment the Sunday offerings. Officers of the 1940s Milanville Episcopal Methodist Church group are, from left to right, Mabel Skinner, Bessie Skinner, Blanche Gabel, and Hattie Kays. (Courtesy of Lavina Jay Powell.)

Three

DAMASCUS

COVERED BRIDGE. In February 1902, an ice gorge swept away the covered bridge, which Solon Chapin had built in 42 days. Another was erected afterwards at a cost of $15,000. It was a three-span structure with two piers in the river, but could not withstand the destructive flood of March 1904. (Courtesy of Albert Noble.)

DAMASCUS SCHOOL FACULTY. Four people comprised the whole staff in the school year of 1932 and 1933. Pictured on the left, from left to right, are John Geist, Mildred Adams, Miss Schuyler, and principal Harry Doll. In the school year 1946 and 1947, the Damascus School had a faculty of nine members. Pictured below from left to right are (first row) Miss Everts, Lucille Rutledge, Gertrude Bortree, and Mildred Adams; (second row) Dorothy Krause, Edgar Jenkins, Mr. Kent, Mr. Stormer, and LaRue Elmore. (Courtesy of the Damascus Township Historical Society.)

HIGH SCHOOL SPORTS CLUB. Damascus High School was very active in sports, as noted in this 1944 photograph. Seen here from left to right are (first row) Francis Wilcox, Alice Hagenow, Jean Jackson, and Rudy Durcansky; (second row) Virginia Dutton, Beverly Orr, Juanita Hocker, Betty Coe, Irene Tegeler, Marian Davis, and Avis Smith; (third row) Robert Diehl, Milton Vail, Donald Rutledge, Leonard Rolston, Victor Dennis, Robert Sheard, and Nelson Plain. (Courtesy of the Rolston family.)

Eighth-Grade Graduation. Celebrating their achievement, the class of 1952 was entertained Western-style. The performers are Roger Swendsen on the left, and John Eldred. (Courtesy of Roger Swendsen.)

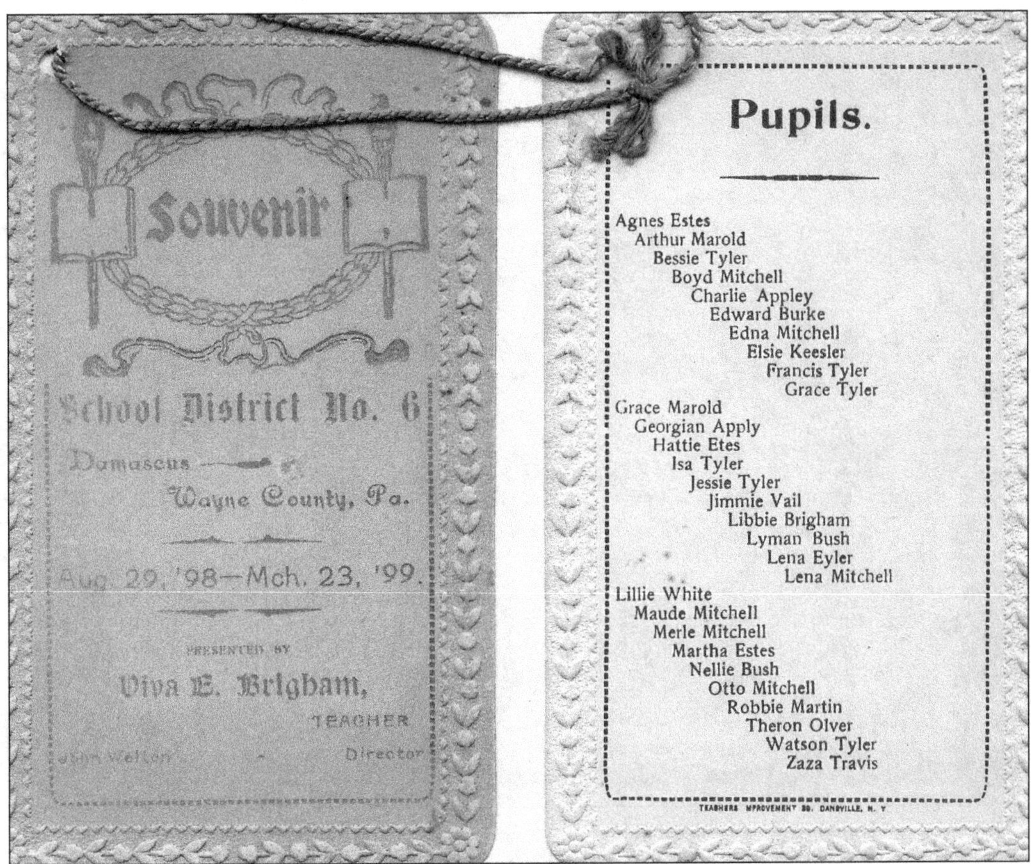

Souvenir

School District No. 6

Damascus

Wayne County, Pa.

Aug. 29, '98—Mch. 23, '99.

PRESENTED BY

Viva E. Brigham,

TEACHER

John Welton - Directors

Pupils.

Agnes Estes
Arthur Marold
Bessie Tyler
Boyd Mitchell
Charlie Appley
Edward Burke
Edna Mitchell
Elsie Keesler
Francis Tyler
Grace Tyler
Grace Marold
Georgian Apply
Hattie Etes
Isa Tyler
Jessie Tyler
Jimmie Vail
Libbie Brigham
Lyman Bush
Lena Eyler
Lena Mitchell
Lillie White
Maude Mitchell
Merle Mitchell
Martha Estes
Nellie Bush
Otto Mitchell
Robbie Martin
Theron Olver
Watson Tyler
Zaza Travis

TEACHERS IMPROVEMENT CO. DANSVILLE, N. Y.

DAMASCUS SCHOOL NUMBER SIX. At this Damascus school, there were 30 students attending during the six-month school term, which ran from August 29, 1898, to March 23, 1899. Pupils listed are Agnes Estes, Arthur Marold, Bessie Tyler, Boyd Mitchell, Charlie Appley, Edward Burke, Edna Mitchell, Elsie Keesler, Francis Tyler, Grace Tyler, Grace Marold, Georgian Apply, Hattie Estes, Isa Tyler, Jessie Tyler, Jimmie Vail, Libbie Brigham, Lyman Bush, Lena Eyler, Lena Mitchell, Lillie White, Maude Mitchell, Merle Mitchell, Martha Estes, Nellie Bush, Otto Mitchell, Robbie Martin, Theron Olver, Watson Tyler, and Zaza Travis. Viva E. Brigham was the teacher, and John Welton was the director. (Courtesy of the Damascus Township Historical Society.)

UNION ACADEMY, C. 1881. Chartered in 1849, the academy building was constructed at a cost of $1,050. Financial troubles followed until 1879, when a new board of trustees renovated the building with gains realized from lumbering off the academy property. The remodeled school was 72 feet long, 24 feet wide, two stories, and featured a cupola 65 feet high. In 1899, the academy was sold to the school directors for use by Damascus Township, with a three-year high school course of study being introduced. Beginning in the fall session of 1909, a four-year high school study began. (Courtesy of Helen Dexter.)

Rafting on the Delaware. In 1764, Daniel Skinner constructed and navigated the first raft down the Delaware River. The *Newton Register* reported, "In 1859, the largest one was 195' long by 65' wide and required a crew of eight men. Included in, and on the raft were 3,000 railroad ties, 6,000 hoop poles, 32,000' of joists, 108,000' of wharf timber, 2,500' of timber, and 10 ship knees." The last of rafting on the Delaware took place around the 1920s. Pictured below are two hoop bows used in lashing a raft together. The background of the photograph was taken from the 1882 diary of Damascus resident J.E. Brush. Journal entries earlier that year show that from February 13 to March 2, he made 2,963 bows and sold them in Milanville on March 3. (Above, courtesy of Roger Swendsen; below, courtesy of the Damascus Township Historical Society.)

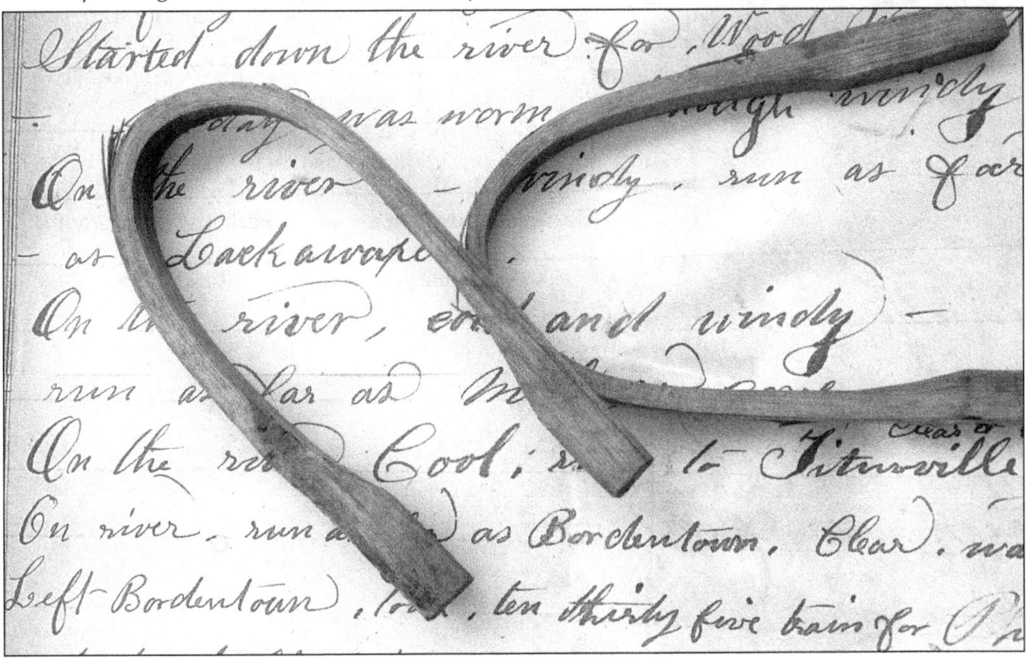
Started down the river for Wood

was warm

On the river – windy. run as

as Lackawa

On the river, and windy –

run as far as

On the Cool; to Titusville

On river. run as Bordentown. Clear.

Left Bordentown, ten thirty five trains for

Basketball Uniforms, 1917–1918. In these two photographs, there is a stark difference between the uniforms of the boys and girls teams. The girls had knee socks, loose-fitting knickers, and navy-style shirts, while the boys were sporting striped, rolled-down knee socks, shorts, modified tee shirts, and two-toned court shoes. The girls ended the season as basketball champions, and the boys, style kings. (Courtesy of the Damascus Township Historical Society.)

Damascus, Class of 1910. The high school was now a four-year course of study for these students. Pictured from left to right are (first row) Orville Brigham, Edward Fromer, John Stephenson, Russell Clark, C. L. Tegeler, Robert Boyd, Vernon Tegeler, and Orville Welsh; (second row) Eva Snavely, Ava Reynolds, unidentified, Helene Yerkes, Juanita Branning, Nina Smith, Elsie Gregg, Mary Loy, ? Ostrander, Anita Skillhorn, and Grace Stephenson; (third row) Merritt S. VanCampen (principal), Mary Noble, Ruth Kopp, Frederica Hocker, Helen Rutledge, Clara Gaston, Mary Fromer, Sadie Welsh, unidentified, unidentified, Florence Sheard, and Carrie Snaveley; (fourth row) Mabel Reynolds, Madelyn Branning, Alma Canfield, Myrtle Reynolds, Susie Mosher, Lovisa Sheard, Alma Noble (teacher), John Burke, Merle Bogert, Guy Reilly, Theron Lillie, and teacher Walter Sheard; (fifth row) teacher Spencer Noble and Frank Reilly. (Courtesy of the Damascus Township Historical Society.)

Sugarhouse. Built in the late 1800s, this sugarhouse had been in the Canfield family and used consistently until the late 1940s. Evaporation of the maple sap from the 40-gallon, seven-foot long flat pan took place over a wood-fired bed of coals. (Courtesy of Karl Canfield.)

Champs of 1932. Even 80 years ago, the girls' basketball team at Damascus High School were at the top of their game. Pictured here from left to right are (first row) Bertha White, Lucille Rickrich, and Laura Hinnaman; (second row) Hildreth Doll, Gussie Markschied, and Dorothy Farley. (Courtesy of the Damascus Township Historical Society.)

Main Street in Damascus. In the background of the above photograph is the river bridge surrounded by floodwaters. Stability of the bridge at such a time was a concern, since the nearest crossing was miles in either direction. On the left are onlookers at the steps of the Knapp and Theobald store. In 1895, Edson Knapp purchased the old Vail and Appley store, which had been vacant since 1873. He remodeled the building, and within a year's time, the business grew and he had to enlarge the store. In June 1896, he married Gertrude Jackson of Laurel Lake, and they resided in an apartment over the store. Henry Theobald became a business partner in 1899, and the enterprise prospered even more than before. Taking a wintery ride in the cutter below are Edson and Gertrude Knapp. (Above, courtesy of the Damascus Township Historical Society; below, courtesy of Opal Hocker.)

The General Store Keeper. As in most general stores, the keeper was ready to fill requests from the variety of goods offered. At the right is a humidor filled with cigars; on the next counter are rolls of butcher paper and twine hanging down, handy for wrapping meats; on the floor are sacks of flour; and the shelves are stocked with canned goods and supplies. Note that near the centrally located stove is a coal-filled wooden bucket with "Carbondale, PA" on it. (Courtesy of the Damascus Township Historical Society.)

Methodist Church. The First Methodist Episcopal Church of Damascus was dedicated in October 1857 with less than a dozen members. Total cost of land, church, and furnishings was just over $1,600. Work performed in 1874 and 1875 was dedicated to the enlargement and remodeling of the church, as well as erecting a steeple. The picket fencing around the churchyard was necessary to keep roving farm animals from trampling the grounds. (Courtesy of Roger Swendsen.)

Delaware Valley Motor Company. Historians relate that the center part of this building was a church that had been moved from near Card's flats at Cushetunk to the foot of the hill in Damascus. Later, the sides with sloped roofs were added on, and it was used as a feed store before being a garage. Thomas Fortnam Jackson, born in 1873, was the owner of the Delaware Valley Motor Company at the time of this photograph. Thomas Griffith and his son Dwight also sold Ford products in Damascus during the 1920s. Partially assembled Ford automobiles would arrive by train at the Cochecton Station, with final assembly of fenders, running boards, and such taking place in Damascus. When the new river bridge was constructed in the 1940s, the building was moved once again, but just back from the road. The business, last known as the Damascus Garage, closed in the late 1980s with Charles Grady as owner. (Courtesy of Opal Hocker.)

The Paint Mine. When Nathan Mitchell settled in Damascus, he befriended some Indians. They showed Mitchell and his sons where a soft rock, or red clay, was located. The boys pulverized the rock and found that when mixed with buttermilk, it produced a wonderful barn paint. Tons of this pigment were shipped for paint purposes, and the Mitchell family was known for having the best barn paint in the territory. (Courtesy of the Damascus Township Historical Society.)

Oak Grove Cottage. A newspaper clipping from November 23, 1905, described this building before it was constructed: "Miss Ellen McGue is having an elegant residence erected on the hill overlooking the Delaware River. John Wood of Atco is the builder. The house is very large and has all modern conveniences. It will be opened in the spring." The boardinghouse was located on Brucher Road between Milanville and Damascus. (Courtesy of Helen Dexter.)

Damascus High School. To accommodate an increase in high school enrollment, the academy was replaced with a two-story brick building in 1929. With six large classrooms, a stage in the gymnasium, locker room and showers, an office, sick room, storage areas, lavatories, and other amenities, the school was an educational facility of which Damascus Township was proud. Members of the board, staff, and community are assembled here at the dedication. (Courtesy of the Damascus Township Historical Society.)

Eighth-Grade Graduation. "At the top but climbing," was the motto of this 1957 class at Damascus. The happy students from left to right are (first row) unidentified, Andrew Vanatta, Gary Hocker, Ernest Rutledge, Linwood Duke, Alice Welsh, Marion Rutledge, Charles Grady, Carolus Davis, and April Olver; (second row) Joyce Bridges, Virginia Bortree, unidentified, Raymond Mohn, Wayne Ellison, George Keesler, John Diehl, unidentified, Richard Pingel, Judy Swendsen, and LaVon Canfield. (Courtesy of the Damascus Township Historical Society.)

Baptist Church. The Baptist Church in Damascus was organized by 12 people and recognized on August 26, 1821. Rev. John Smitzel was the first pastor, dividing his time between the churches in Damascus, Bethany, and Clinton Center. The first church building was erected by Thomas Shields and was in use until 1831. Afterward, a new church was built on the site by members. (Courtesy of Helen Dexter.)

Canfield Farmstead. Isaac Canfield built this home for his wife, Hattie Brooks, and their five children in the late 1800s. A unique feature of this eight-room house was the port for the horse and carriage in inclement weather. In use until 1982, the farmstead saw 18 children raised through four generations. (Courtesy of Karl Canfield.)

HUNTER SAFETY CLASS. In 1959, Pennsylvania offered voluntary safety instruction for the first time. This class, held at Damascus High School in 1961, certified 29 students. Pictured from left to right are (first row) Myra Schnakenberg, Kathy Ostrander, Kathy Griemsman, Susan Jay, Peggy Welsh, Carl Bell (instructor), Mary Keesler, Rosemary Herold, Mark Olver, Wayne Lawrence, James Crum; (second row) Thomas Meehan (District Game Protector), Howard Welsh, Edwin Tyler, John Yates, Mike Ritch, David Mitchell, Walter Hull, Charles Sutliff, Phillip Keesler, Kenneth Tyler, Fred Weigelt (District Game Protector); (third row) Charles Heyn, James Seipp, James Yanacek, Donald Brown, Schuyler Dexter, Joseph Davis, LeRoy Canfield, Laird Ritch, Joseph Blaine, and Gerald Brown. (Courtesy of the Damascus Township Historical Society.)

Doctor's Office. Dr. Theron Appley and his son Dr. Otto Appley used this c. 1850 building as their office for seeing patients. Currently, the Damascus Township Historical Society occupies it, maintaining an archive museum with displays of artifacts, as well as reference and research items. (Courtesy of the Damascus Township Historical Society.)

The Iron Bridge across the River. A local news item in 1920 stated, "traffic halted on the Damascus-Cochecton Bridge. A five-ton Packard truck loaded too high with barreled apples broke through the planking at different places and at last went through to the hub of one wheel. The bridge tender warned the driver he thought the load exceeded the capacity of the bridge. It is reported the truck was on its way to New York City from the orchard of Henry Heins at Milanville Heights with 70 barrels of apples." (Courtesy of Helen Dexter.)

GIRLS BASKETBALL TEAM, 1946–1947. Winning 15 out of 18 games that season, the Damascus High School girls' team was ranked as one of the top in Wayne County. Veramae Monington scored 117 points and Luella Coe had 101. From left to right, they are (first row) Juanita Hocker, Anna Mary Noble, Marie Hopkins, Luella Coe, Beulah Dutton, and Veramae Monington; (second row) Barbara Testa, Marilyn Vail, Irene Mach, Betty Vannatta, coach Lucille Rutledge, Margaret King, and Lois Diehl; (third row) Lillian Goodenough, Peggy Peck, Margaret Moore, and Fay Newenhouse. (Courtesy of the Damascus Township Historical Society.)

CHEERLEADERS. Damascus High School had their very first cheerleaders for their basketball team in 1947. Pictured from left to right, the troupe of three consisted of Irene Mach, Veramae Monington, and Marie Hopkins. (Courtesy of Ralph Smith.)

Graduating Class of 1940. Seniors at Damascus High School had a custom of holding class night where a skit was performed and fun awards were presented. The Western-style classmates from left to right are (first row) Shirley Schweighofer, Esther Highhouse, Iva Miller, Mary Smith, Doris Hankins, Thelma Theiss, Wanda Abraham, Alice Branning, Eleanor Rolston, Marge Edwards, Roberta Keesler, Phyliss Geiss, and Eleanor Noble; (second row) Frank Vondrich, Ulmont Dexter, Martin Dermody, Sherman Erich, Wyman Rutledge, Junior Erich, Fred Tegeler, Grace Smith, Vivian Thomas, Mavis Gager, and Evelyn Terrell. (Courtesy of Ralph Smith.)

Four

Tyler Hill

Andrew and Maria Swendsen Family. Andrew Swendsen arrived in America in 1869 from Denmark. Three years later, he sent money for his betrothed, Maria Vinding, to join him. As soon as she arrived June 1872, they married at the Methodist parsonage. Pictured here with their nine children around 1900 are, from left to right, (first row) Martha, Andrew, Mary, Maria, Peter, and Anna; (second row) Dora, John, Adam, Frank, and Fred. (Courtesy of Roger Swendsen.)

Tyler Hill Post Office. When the post office was first established in Tyler Hill in 1878 with David Fortnam as first postmaster, it was located in this store, owned by Fortnam and his stepson William A. Smith. Upon closing of the business in 1907, the store goods were moved and the post office was relocated to Clarence Pethick's store until 1933. On the left side of the porch is Fred Coe, and seated by the open door is Clarence Pethick, Fortnam's son-in-law. (Courtesy of Jacque Rutledge.)

Pethick's Tyler Hill Store. Israel Tyler's sons, L.D. Tyler and Moses Tyler, built the original, westerly part of the store. Clarence M. Pethick had a large addition built on the eastern side in 1908. For many years, the upstairs of this addition was used for the Dairymen's League and other meetings, and also as a polling place for voting. (Courtesy of Helen Dexter.)

Andrew Swendsen House. This scene was the result of a rare winter lightning storm on January 8, 1907, at the home of Maria and Andrew Swendsen in Frosty Hollow. She had just left the living room where her granddaughter Elsie Jensen remained at the time of the strike. The *Wayne Independent* reported at the time, "The doors had been torn from their hinges, every pane of glass shattered, the sitting room heater and a radiator were in melted fragments, the living room floor had been broken with a gaping hole into the cellar, wires in and about the house were melted, a pound and a half of blasting powder in the kitchen had exploded, the earth in several directions was furrowed. Andrew and his son, Jacob, who were approaching the house at a distance of about 200 feet, were thrown violently to the ground." All miraculously escaped injury. (Courtesy of Roger Swendsen.)

Hay Wagon. Pictured atop this wagon full of hay is Clarence Stalker by the Giles's farm near West Damascus. To the left is a hay loader, which made work easier than loading by hand onto the wagon. (Courtesy of Doris Bryant.)

Stone Arched Bridge. Located in Tyler Hill, most of this hand-laid stone arched bridge was swept away in the "pumpkin" flood of October 1903. The nickname "pumpkin" was attached to any flood that occurred during the time of pumpkin harvesting. (Courtesy of Roger Swendsen.)

One-Room Schoolhouse. Brothers John and Richard Olver built the Tyler Hill School in 1875 for $1,000. Throughout the years, it had many uses besides educational, with a farmers' institute, Dairymen's League meetings, Sunday school, and church services also held there. Cornelia Bryce Pinchot, wife of the former governor of Pennsylvania, gave a political talk there when she was running for Congress in April 1928. Pictured above in 1920 are Ivan Swendsen (left) and Lon Ellison. Attendees below, pictured around 1937, are, from left to right (first row) Keith Swendsen, and Gerald Dutton; (second row) Gideon Theis, Richard Dutton, Harry Smith Jr., Beulah Dutton, Virginia Hlavac, and Barbara Eby; (third row) Nelson Hinaman, Arnold Dutton, Jean Jackson, Gladys Welsh, and Jean White; (fourth row) Wesley Weber, Nora Hinaman, Pearle Hinaman, Doris Welsh, and teacher Inez Hocker. The school closed in 1962. (Courtesy of Roger Swendsen.)

Mowing Machine. Ivan Swendsen is mowing hay in 1930 with Frank Swendsen's team of horses, Bill (on the left) and Prince. (Courtesy of Roger Swendsen.)

Jacob and Abigail Boults Lewis House. First cousins Fanny and Gertrude Rolston Swendsen are standing in front of their grandparents' home in the early 1960s. It was where their mothers, Laila and Octavia, were born. (Courtesy of Roger Swendsen.)

Wayne Country Club. In 1922, brothers-in-law Morris Kamber and William Mitchell purchased the Will Jackson farm on the west side of Laurel Lake, built a golf course, and started what became the Wayne Country Club. In 1923, they decided to each go their own way, with Kamber retaining the country club and Mitchell keeping the camp. Kamber's son Bernie was a newspaper columnist, and celebrities, such as Burt Lancaster, Red Buttons, Rita Hayward, Frank Sinatra, Joe DiMaggio, Jackie Gleason, George Burns, and others were noted in his entertainment column as guests at the country club. With the popularity of flying to destinations and the decline in rail passenger travel, the business dwindled, and in 1954 closed operations. It went into bankruptcy in 1955 and became Tyler Hill Camp in 1956. (Courtesy of Jacque Rutledge.)

Silver Lake House. This boardinghouse of eight rooms and 10 cabins was owned and operated by Edith and Norman Dennis from 1951 until 1976. Filled mostly with guests from New York City, it had been featured as a "farm vacation" with activities like horseback riding, fishing, boating, and swimming. Even with a full house of 70, Edith did all the cooking, including homemade pies. Originally, the farm was owned by Norman's grandfather, Jacob Welsh. (Courtesy of Evelyn McGowan.)

Winter 1945. As was the case all over the northeast that year, snow and wind made travel treacherous, besides having to haul milk to the creameries. Pictured here with his sleigh is Russell Welsh along with the workers who helped haul milk cans to the creamery and bring feed on the return trip. (Courtesy of Doris Bryant.)

Days of Yore. The richness of the earth, fair weather, a team of sturdy workhorses, and a good photographer almost make farming seem like a romantic, if not idyllic, way of life. While the latter is true, there are many hours of toil in all conditions. Pictured working his Tyler Hill farm is Harry Smith Sr. (Courtesy of Ralph Smith.)

Maple Sap Collecting. Frank Swendsen is readying this maple tree for collecting sap. After boring two or three holes in the tree, the spiles would be inserted, and a bucket placed under each. Although metal spiles were available, he made his own from sumac, as the core of sumac is hollow. Five-year-old grandson Roger is his helper. (Courtesy of Roger Swendsen.)

ICE HARVESTING ON LAUREL LAKE. Before the days of modern refrigeration, homes would have iceboxes, which were cabinets where a block of ice would keep foods cool. Harvested from nearby lakes, ponds, and also the Delaware River, the blocks of ice would be shipped to many cities by rail. The c. 1940 photograph on the left looks down from the platform chute, where the water channel with ice blocks can be seen with the ice-cutting machine in the background. Below, pushing ice up the loading ramp are Leroy Gries (left) and Russell Welsh. (Courtesy of Doris Bryant.)

Readying the Field. Russell Welsh is pictured in the late 1930s steering the plow drawn by his team of horses in preparation of planting the field. His farm was located near the top of the Tyler Hill village. (Courtesy of Doris Bryant.)

Camp Mitchell. The first boys' and girls' camps in Wayne County were located on Laurel Lake. Both founded around 1910, the boys' camp, Camp Harlee, and the girls' camp, Camp Mitchell, were owned by Morris Kamber and William Mitchell for many years. An outbreak of polio around the early 1950s took the lives of three campers, along with the camp doctor. (Courtesy of Roger Swendsen.)

Air Force Plane Crash. In September 1949, this Air Force cargo plane crashed on the Norman Dennis farm, not far from Damascus and Lebanon Townships' line. When the pilot became aware of trouble, he radioed back to the three passengers in the cargo hold to bail out. The pilot did so, landing in a tree. He suffered a broken leg and was taken to the Dennis home for transport to a hospital. There were three mortal casualties in the wreckage—the servicemen had been asleep, and did not hear the pilot's warning. (Courtesy of Marie and Robert Diehl.)

Sawing Wood. In this photograph, it appears that five-year-old Roger Swendsen is being of more help than when collecting maple sap. The saw he and his grandfather Frank are using is a two-man crosscut saw, common on every homestead in the era. (Courtesy of Roger Swendsen.)

THE OLD MILL. This mill was located near Tyler Hill and was typical of most early 1900s sawmills. The *Sullivan County Democrat* reports, "May 28, 1912, William Smith purchased the Dobson steam mill and all its fixtures. His will supplant Mr. Smith's water powered saw mill," and, "March 18, 1913, William Smith has begun operation in his new steam sawmill. Mr. Applegate is running the saw, and Isaac Warner is attending the engine." The sawmill Smith owned was located on Hopkins Road. (Courtesy of Jacque Rutledge.)

CUTTING CORN. As was the custom with most early farmers, one of the most important things he owned was a good team of horses. In the 1930s, Dwight Griffin is using this team for cutting corn. Note the fly nets covering the horses. In 1895, the Montgomery Ward Catalogue sold a leather team net with 60 lashes for $2. (Courtesy of Doris Bryant.)

At the Well. A common sight in rural areas was the hand pump, as many homes did not have running water until as late as the 1950s. Even the youngest of children knew to leave a bucket of water at the well for priming the pump the next time water was needed. Pictured here pumping water is Harry Smith, Jr. (Courtesy of Ralph Smith.)

Tyler Hill House. William A. Smith was known by the nickname "Basswood Billy" to differentiate from the Tyler Hill mill owner, William Smith. He earned the moniker because he had purchased a large amount of basswood from the Lovelass Brothers' sawmill in Fallsdale to build the Tyler Hill House and then neglected to pay for it. In 1881, the sheriff's sale of the house also included the store next door that he had owned with his father-in-law, David Fortnam. Later owned by Charles Schulmbaum, it was a thriving boardinghouse for many years. (Courtesy of Jacque Rutledge.)

Five

THE NORTHERN TIER

CONKLIN HILL UNION CHURCH. In 1913, William Boults gave one-third of an acre from his farm parcel to the trustees of the church. With E.D. Bennett as the architect and a Mr. Lowe of Callicoon as master builder, the structure was finished in 1915. Attending this 1932 Sunday school class from left to right are (first row) Virginia Keesler, Doris Mitchell, Ruth Wheelock, Mary Keesler, and Wanda Monington; (second row) teacher Bessie Whitmore, Jesse Conklin, Ralph Hull, unidentified, Vincent Mitchell, James Hull, Donald Turner, and William Smith; (third row) unidentified, Raymond Turner, Fred Marold, and Orville Whitmore. (Courtesy of Opal Welk Hocker.)

Conklin Hill School. Pictured in this pre-1932 photograph from left to right are (first row) Alberta Bush, Millie Grena, Mary Keesler, Harold Turner, Roberta Keesler, Erma Monington, and Raymond Deighton; (second row) Virginia Keesler, Anna Grena, Ruth Wheelock, Zaza Turner, Wanda Monington, and Stanley Deighton. Note the many cords of firewood behind the children, as a wood stove was used to heat this one-room schoolhouse. (Courtesy of Opal Welk Hocker.)

Frank Rutledge's General Store. Centrally located in Galilee, the business had been home to the telephone switchboard until January 1910, when the Big Eddy Telephone Company moved it to the residence of Bertha Rutledge. A new feature of gasoline pumps was added to the store on June 1, 1915, even though travel by horse was more common than by automobile. (Courtesy of Helen Dexter.)

A.J. ABRAHAM AND SON. At their general store, everything from general merchandise, furnaces, International machinery, feed, seeds, and eggs was sold. Writer J.T. Bradley stated, "In 1913 when the egg rush is on, 10 to 15 cases a day are at the store. Last week a team loaded with 214 empty cases was on the way to the store, returning from the rail station at Cochecton." (Courtesy of Helen Dexter.)

BARREL FACTORY. George C. Abraham was an enterprising fellow. Not only did he own a creamery and cheese factory, but in 1904 he also built and started work putting up apple barrels in his cooperage. In the 1930s, a dance hall was located on the second floor of the building, and later a grange store operated from it. One of the last owners was Clarence Schwab, who had a garage there with living quarters upstairs. (Courtesy of Jacque Rutledge.)

CONCRETE HOUSE. Frank Newberger of Jeffersonville, New York, built the new dwelling of Agnes Smith in 1908 with his patented concrete blocks. Her home was also her place of business, as this was where she ran a millinery shop, the telephone central, and later, an ice cream shop. In 1908, telephone poles were set in Galilee, with service having 24 numbers on a party line. Across the street from her home was the post office, run by Frank Tyler. Local lore has it that since her parents were not fond of him, there could be no courting. However, at the age of 76, Agnes Smith and Frank Tyler were married. (Courtesy of Helen Adams.)

CHURCH STREET. Up from the four corners in Galilee is the Methodist church. Organized in 1840 by Joseph Sutliff, they held services at a nearby schoolhouse for 35 years until the Galilee Centennial Methodist Church was granted a charter in December 1875. With land purchased from Sutliff and supplies donated by members, the building was erected in 1876. (Courtesy of the Wayne County Historical Society.)

Abrahamsville Church. The church was just over 10 years old when this Sunday school class assembled on September 1, 1935, to begin a new season. Ground-breaking for the church had taken place July 1925 with Judson Canfield as master builder on land deeded for the church by Henry "Joe" Whitmore. From left to right are (first row) teacher Mrs. Clinton, Clinton Henkie, Orvis Whitmore, Nelson Miller, and James Baker; (second row) DeWitt Starkie, Margaret Chilson, Kay Chilson, Beatrice Martin, Iva Miller, and unidentified; (third row) Jack Paulson, Alma Lord, Martha Paulson, Nellie Starkie, Elsie Walker, Alice Pangburn, and Vera Baker. Since it was not a custom for local girls to wear hats, the "hatted" ones were "city girls." (Courtesy of Brian Rutledge.)

Abrahamsville School. In May 1926, the students attending school are, from left to right, (first row) James Baker, Nelson Miller, Thomas Litzenbauer, Alma Lord, Vera Baker, Emily Blair Evans, and teacher Olive Calder; (second row) unidentified, Everett Schuman, Ebenezer Essex Keesler, Louise Walker, Mary Whitmore, Elsie Walker, and Lillie Keesler; (third row) William Knospler, possibly Everett Keesler, Bertha Baker, and unidentified. (Courtesy of Brian Rutledge.)

Marks Family. The surname of Marks is an old, established name in Abrahamsville. Generations of Jacob Marks, like his son Jacob and grandson Frank (born in 1872), worked the family farm. From left to right are Frank Austin Marks, May Marks Stalker, Amy Beltz Marks, Harry Stalker (child), John Beltz, Martha Hill Marks, and Harry Stalker. (Courtesy of the Marks family.)

The Henhouse. In the late 1930s, a two-level chicken coop was built by Emry Marks, son of Frank Marks, and in the mid-1940s, the four-story section was added. Eggs from 4,000 chickens were sold to individuals as well as local establishments. Bagged feed was also sold from this building until the bulk feed industry took over. (Courtesy of the Marks family.)

Rutledgedale School. Although the number attending the one-room schoolhouse was small, the curriculum was just as rigorous as the larger schools. More than likely, students advanced more quickly due to the availability of individual attention. Pictured in this 1913–1914 class photograph are, left to right, (first row) Margaret Steffen, Harold Rutledge, and Grace McDivitt; (second row) Wayne Rutledge, and Anna Steffen; (third row) Beatrice Rutledge, teacher Florence Wood, and Henry Steffen. (Courtesy of the Damascus Township Historical Society.)

Lookout Church. In 1890, the piece of land sold for $1 by Mr. and Mrs. Charles Fulkerson to the trustees of the Methodist congregation had a stipulation. The requirement was that a church had to be erected within three years. In just a little more than one year of the land sale, the church was dedicated on November 26, 1891. To the right of the church is a carriage barn with 10 stalls. On the far right is the parsonage. (Courtesy of Myra Schnakenberg.)

Wash Day. With a boardinghouse full of guests, nearly every sunny day was wash day. In the 1914 photograph on the left, Gus Schnakenberg, about age 12, is working the basket of the washing machine. The slotted basket, made of woven wood, was rocked back and forth through the water in the tub. The tub part was wood and metal, and the frame of the machine was wood. Below, the woman on the left is hand scrubbing while the other is wringing out excess water before hanging items on the line to dry. (Courtesy of Myra Schnakenberg.)

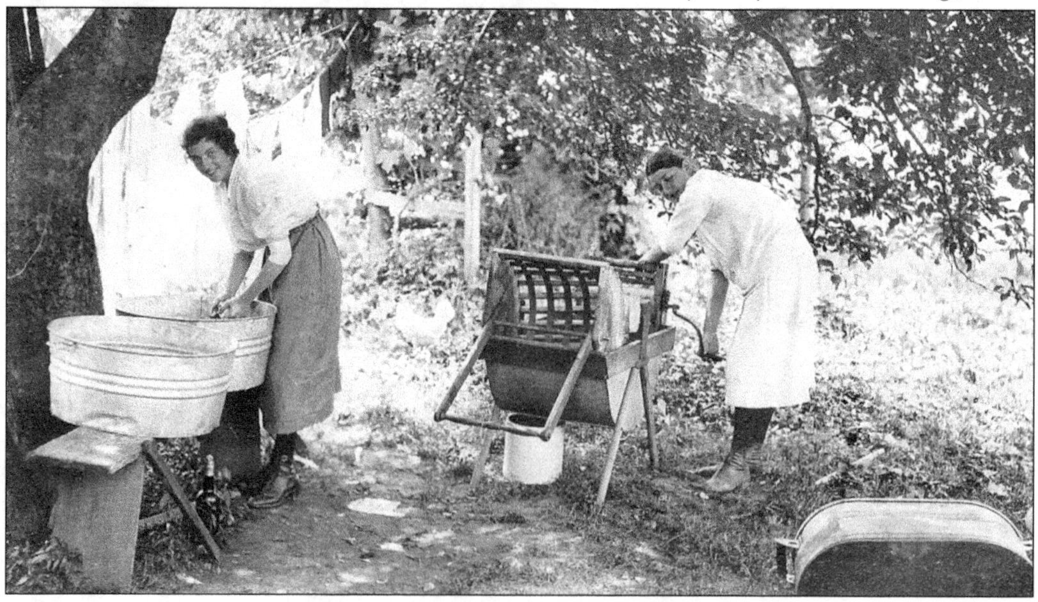

LOADING HAY. Before the days of automation, the word "farmhand" could be taken quite literally. On the Schnakenberg farm, these farmhands were intent on getting the job done, while the team of horses waited patiently for the flurry of activity to cease so they could head for the barn. (Courtesy of Myra Schnakenberg.)

ICE CUTTING. Joel Hill's gristmill pond was the location of this crew of ice harvesters. The pond was located opposite the church in Lookout. Pictured from left to right are Clarence Schwab, unidentified, William Schnakenberg, Glenn Schwab Sr., Nelson Hill, and Gus Schnakenberg. (Courtesy of Myra Schnakenberg.)

Village of Hilltown. Upon Joel G. Hill's return from the Civil War, he was employed as a sawyer at William Holbert's and J.D. Branning's Beaver Meadow mill in Lookout. A few years later, he worked at Isaac Young's sawmill. In 1876, he purchased Young's sawmill and acreage, and in 1898, he bought the former Holbert and Branning lands along with the sawmill and large

pond at Duck Harbor. In 1902, Joel G. Hill purchased 2100 acres from Isaac Young and Company. With the tremendous amount of land and business holdings of Joel Hill, the area was unofficially known as Hilltown. (Courtesy of the Wayne County Historical Society.)

Hill's Acid Factory: With all the virgin timber cut by 1898, what remained was ideal for use by the acid factories. In order to build his own factory in Lookout, Joel Hill borrowed money from William Riefler. Building the acid factory required much from the local lumber mills because of its sheer size. The *Honesdale Citizen* of May 10, 1900, wrote, "Philip Van Orden is doing the carpentry work at Joel Hill's saw mill at Duck Harbor, and preparations are being made to saw out the frame timber for the acid factory at Lookout." The business closed in 1935, long after his death in 1919. Pictured below is Hill's home at the top left, and at the rear of the apple orchard is Hick's store. (Courtesy of the Equinunk Historical Society.)

Joel Hill Sawmill. This mill, built by William Holbert and J.D. Branning soon after the Civil War, was purchased by Joel G. Hill in 1898. A waterwheel had powered the mill until the "pumpkin" flood of 1903, when it was washed away. The mill continued to operate until the early 1970s, with Nelson Hill and Joel G. Hill II as sawyers. In 1974, it was added to the National Historic Sites Register. After being donated to the Equinunk Historical Society in 1988, operational restoration commenced. It is the only water-powered sawmill remaining in northeastern Pennsylvania. (Courtesy of the Equinunk Historical Society.)

Youngsville School. During the 1955 school year, the Youngsville one-room schoolhouse had 27 pupils in grades one through seven, all taught by Elizabeth Rutledge. Some of the students pictured, in no particular order, are Emmy Lou Hill, Virginia Bortree, Raymond Bortree, Lindsey Cole, Ronald Cron, Judy Gaunt, Myra Schnakenberg, Carol Hoehn, Nora Rutledge, Marian Rutledge, Gary Rutledge, Nelson Hinkley, Richard Rutledge, Harold Rutledge, Walter Kellam, Sandra Meyer, Darlene Chitman, and Eloise Hinkley. (Courtesy of Myra Schnakenberg.)

WELSH FARMS CREAMERY. This facility was a milk receiving station for milk cans. Its operation started on August 23, 1946, with 13 patrons. In the summer of 1962, the conversion to bulk tank took place, and then they had a total of 60 patrons. The business ceased in 1976, 30 years after it began, and the building was sold. (Courtesy of Harold Hawley.)

HARVEST TIME. The piece of machinery in this photograph is a self-rake reaper, which Cyrus McCormick made famous in his Daisy reaper. It had a series of arms that swept the cut grain from the quarter-circle platform into gavels, which then had to be hand tied. (Courtesy of Myra Schnakenberg.)

DUCK HARBOR LAKE FARM HOUSE, C. 1910. Herman and Anna Geischen Schnakenberg owned this Lookout-area boardinghouse, which was in operation for several years. Up to 12 guests could be accommodated. (Courtesy of Myra Schnakenberg.)

LOOKOUT GROVE. Located behind the old grange hall, the grove was comprised of elm trees and was the setting for many chicken and biscuit dinners in the early part of the 1900s. It was quite the social affair—not only were people donning their best, but their horses and buggies were a fine sight as well. The milliners must have had quite a trade, as not a bare head is seen in the photograph. (Courtesy of the Equinunk Historical Society.)

Six

Hobbies and Recreation

Tyler Hill Band. In the early 1900s, a common form of entertainment at events was the local band. Assembled here from left to right are (first row) Fred Coe, Scott Rutledge, Arnold Rutledge, and Lloyd Welsh; (second row) Fred Schwab, Corwin Valentine, Robert Mitchell, Selah Olver, Earl Tyler, and Joseph Sheard; (third row) Perry Griffith, Charles Welsh, William B. Yerkes, Ralph Tegeler, and Ned Griffith. (Courtesy of the Damascus Township Historical Society.)

Racehorse and Sulky. Before the automobile age, horses were used for everything from transportation and work to entertainment. This stunning filly is about to go racing with her owner, Russell Dexter, at the reins. With his blacksmith shop on the family farm, he was able to forge whatever shoeing and farrier product his various horses needed. (Courtesy of the Dexter family.)

Hank Merkley. The wilderness around the Old Brook Road in Milanville where Hank Merkley lived was bountiful with all sorts of game. The proud hunter is pictured here at the general store with a prize bobcat that he had shot. Merkley was a congenial sort, ready for fun. At a homecoming skimmelton for Frank and Daisy Medley Davis in 1939, he was launched by friends through an open window into the newlyweds' bedroom! (Courtesy of Joseph Davis.)

Early Scouting. From the salute at the brim of his hat to the tip of his polished shoes, Harry Smith Jr. was a model Boy Scout. Early outfits of the Scouts were copies of the US Army uniforms of that period, and the Boys Scouts of America was the only group granted this privilege. The campaign hat seen here in the late 1930s was in use until 1943, when the garrison cap replaced it. (Courtesy of Ralph Smith.)

Galilee Baseball Team, 1898 to 1903. This photograph is of local historical importance, for Sidney Tyler was the famed photographer. His talent for tonal qualities and composure of subjects was unrivaled. From left to right are (first row), Tom Sutliff, Frank Tyler, Leo Burke, Tom Gilroy, George Sutliff, and Sidney Tyler; (second row), Ezra Potter, James Burke, Steve Burke, Raymond Pethick, and manager William Theodore. (Courtesy of the Damascus Township Historical Society.)

HAM RADIO. Duane Abraham was one of the very first in the local area to have a short-wave radio. He had a room dedicated to this hobby, as it appears the sheer size of the apparatus would require a large space. (Courtesy of Roger Swendsen.)

WINTER TRANSPORTATION. Carl Brucher is seen here riding in a cutter on the family farm in Milanville. The preparation of an ox for a life of hard work began early. Starting with light loads while young and working up to wagons and timber, an ox was considered fully trained by about the age of four. A yoked team can pull heavier loads than horses, using their strong necks and shoulders against the yoke. (Courtesy of Ivan Davis.)

RICHARD PLAIN'S BIRTHDAY. Neighborhood children gathered to help their friend celebrate his birthday. Pictured in the late 1940s from left to right are (first row) John Eldred, Richard Plain, Carolyn Grady, Duane Canfield, and Gary Taggart; (second row) Janet Ellison, Shirley Theis, Robert Pingel, John Bult, Egbert Welsh, Roger Swendsen, Ronald Seipp, Reginald Seipp, and James Welsh. (Courtesy of Roger Swendsen.)

BICYCLE DAY. This was not your typical day heading home from the Calkins School. On this special day, June 7, 1951, school had let out for the summer and bicycles were allowed. The usual afternoon activities for the last day of school also included going to the swimming hole and hiking in the woods. The happy lads pictured from left to right are Grant Sheard, Richard VanDeLinde, unidentified, William Dennis, Garry Sheard, Lawrence Ostrander, and unidentified. (Courtesy of Grant Sheard.)

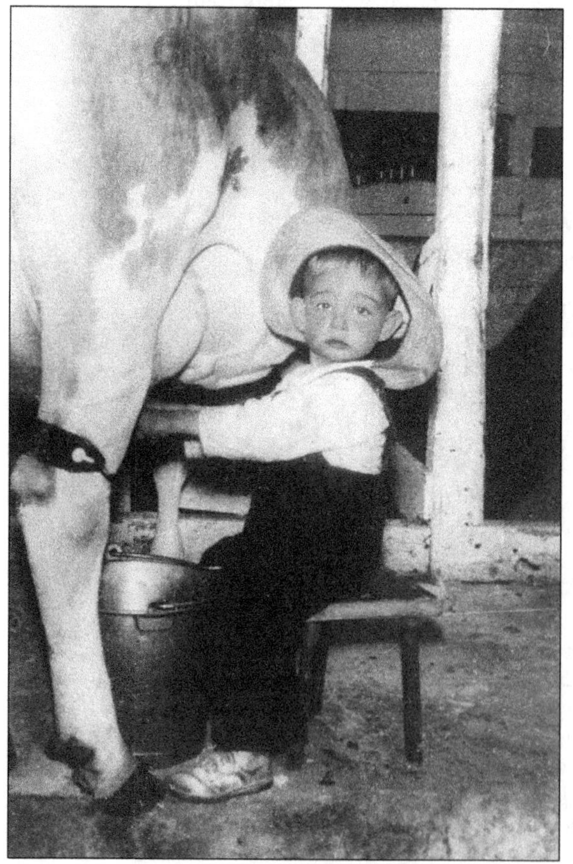

Hunting Party. White-tailed deer season was always anticipated with the hopes of bagging a big buck. This photograph of trophies was taken at the Charles Davis residence in Milanville during the first week of buck season, December 1946. The four Davis men, from left to right, are Frank (8 points, 138 pounds), Edwin (6 points, 122 pounds), Orson (8 points, 142 pounds), Charles (6 points, 126 pounds), and the village barber, Henry Lurtz (8 points, 148 pounds). (Courtesy of the Milanville General Store.)

Tough Times. Putting both hands to work is John Riefler III on the farm of his grandparents, Harry and Florence Davis Smith, around 1950. Even though he was just a tad, it seems that he knew exactly where the milk was and where it should go. (Courtesy of John and Sue Riefler.)

Stalker Reunion. The first reunion of the Stalker family took place at the residence of A. Stalker. On August 31, 1902, 36 people attended the family gathering in Galilee. (Courtesy of Earl Mohn.)

Tyler Hill Eagles Baseball Team. Pictured around 1949 from left to right are (first row) John Bult, Edward Hubert, Charles Williams, Paul Orr, unidentified, and David Orr; (second row) Elvin Swendsen, Robert Lord, Richard Plain, Ronald Seipp, Roger Swendsen, William Lord, unidentified, and coach Nelson Plain. (Courtesy of Roger Swendsen.)

The Damascus-Cochecton Band. This band was comprised of members from two villages across the Delaware River from each other. In existence from the late 1890s until 1910, the band would play at events throughout the area. Pictured from left to right are (first row) George Tyler, Joseph Sheard, Corwin Valentine, Thomas Reilly, and Scott Rutledge; (second row) George Coe, John Sutliff, band leader William Reilly, Kennedy Johnson, Samuel Pethick, and Earl Tyler; (third row) Henry Behrer, Orville Kays, Charles Welsh, Tobias Pethick, Raymond Pethick, Lyman Bush, and Floyd Brigham. (Courtesy of the Damascus Township Historical Society.)

A Successful Hunt. While raccoon fur coats were typically made from trapped pelts, this large raccoon was dispatched by Edward Davis's firearm. Nighttime raccoon hunting, with the use of treeing hounds, was a great sport in the area. (Courtesy of Ivan Davis.)

Swimming Hole. A favorite pastime of locals was taking a dip in Calkins Creek. Enjoying nature's pool near the bridge in Boyds Mills in 1949 are, in no particular order, Jackie Myers, Garry Sheard, Karen Brown, Linda Myers, Grant Sheard, Shirley Myers holding Kent Brown, and Jack Myers holding Kathy Myers. (Courtesy of Grant Sheard.)

OLD-FASHIONED ICE CREAM. Once upon a time, people made all their own ice cream by skimming the cream off the top of the milk, mixing in some eggs, sugar, and vanilla, then putting the mix in the cylinder with ice and rock salt in the keg, and starting to crank. When cranking reached the point of difficulty, it was ready for spoons. Pictured making ice cream is Frank Swendsen of Tyler Hill. (Courtesy of Roger Swendsen.)

AMATEUR BASKETBALL. In the early 1950s, the Damascus Blue Devils were a local basketball team comprised, from left to right, of (first row) Dick Flederbach, Donald Diehl, and Charles Sutliff; (second row) Robert Diehl, Donald Carlson, Donald Orr, and Kenneth Carlson. Robert Diehl started on the varsity team at Damascus High School when he was only in the eighth grade. After graduation in 1944, he played 26 amateur seasons, scoring over 10,000 points during that time span. (Courtesy of Marie and Robert Diehl.)

William Smith's Houseboat. Pictured here with William Smith's prized red oxen is his son Harry Smith Sr. The oxen and cart were used to transport the boat to a launching site. A June 1910 article in the *Sullivan County Democrat* reported, "Several locals took a trip down the Delaware River on a houseboat. They left Damascus Thursday morning and reached Easton, Pennsylvania, Saturday evening, having visited several places of interest along the way, including the Delaware Water Gap. The party reached home Sunday by way of Jersey City, New Jersey. Those who made the trip were Mr. and Mrs. O.W. Brigham, Mr. and Mrs. A.G. Gregg, Charles Mitchell, Sidney Tyler, Mr. Seipp, and Mr. Smith." (Courtesy of Roger Swendsen.)

An Exceptional Trophy. Norman Card and his bobcat are seen on the front steps of the Milanville General Store after the trophy had been measured. While the average is 15 to 20 pounds and about 40 inches long, this particular bobcat weighed 32 pounds and was 59 inches in length. (Courtesy of the Card family.)

Milanville Baseball Team. The Milanville team represented the Milanville Dairy in the Delaware Valley League in 1950. From left to right are (first row) Kenneth Tyler, batboy Paul Orr, John Paparella, and Eugene Paparella; (second row) scorekeeper Jean Mach, coach Albert Henderson Sr., Ivan Davis, William Schutz, and Ed Sobolak; (third row) Phil Sobolak, Percy Tyler, Kenneth Treverton, Robert Diehl, Nelson Plain, and Donald Orr. (Courtesy of the Milanville General Store.)

A Country Gentleman. Irving Rutledge is spiffed up and riding in his carriage drawn by a team of Hamiltonians. He seemed to have thought of everything for a courting date. There is a bouquet of water lilies to his left, a lace-edged lap robe for coziness, and even blinders for the horses! Years later, he traded the horses in for a car. Sidney Tyler of Damascus is believed to have been the photographer. (Courtesy of Brian Rutledge.)

Bush's Glen. Taking in the solitude, beauty, and remoteness of the glen, families would gather here for picnics. In 1758, Simeon Bush came to the Damascus area and was killed by Indians on his way home from Connecticut in 1763. His widow married Bezaleel Tyler, who was killed at the Battle of Minisink in 1779. Later, the Bush and Tyler families established a sawmill on the creek. (Courtesy of the Damascus Township Historical Society.)

A Sunday Afternoon. Seen taking a ride on a Sunday afternoon in this early 1900s photograph are Otto and Louise Snavely Olver. The family farm, located near the pond in Fallsdale, is now owned by a descendant. (Courtesy of the Card family.)

Sledding Fun. In front of the Tyler farmstead was a road ideal for sledding. Winter traffic in the early 1930s was minimal, and with a pony to pull the sled up the steep hill, nothing could be better. Enjoying the snowy day are the Tyler children, Kenneth on the sleigh, and Irene at the front of the pony. (Courtesy of Richard and Nettie Tyler Swendsen.)

AN ANNIVERSARY AT HOME. John and Julia Rutledge Davis observed their 55th wedding anniversary at their Milanville home on October 28, 1956. He had been the township's supervisor for many years in the early 1900s and was well known as an expert woodsman. She played the pump organ for the Milanville Methodist Church and was active in its ladies groups. They had 14 children, four of them dying very young. Children attending the celebration from left to right are Vernon, Florence Smith, Marian Noble, Charles, Ezra, Frank, Alma Mueller, and Russell. (Courtesy of Barbara Davis Dexter.)

www.ingramcontent.com/pod-product-compliance
Lightning Source LLC
LaVergne TN
LVHW081528100826
845153LV00004B/228

* 9 7 8 1 5 3 1 6 6 2 3 0 1 *